Essentials of Safety

Essentials of Safety

Maintaining the Balance

Ian Long

CRC Press

Taylor & Francis Group

Boca Raton London New York

CRC Press is an imprint of the
Taylor & Francis Group, an **informa** business

First edition published 2022

by CRC Press
6000 Broken Sound Parkway NW, Suite 300, Boca Raton, FL 33487-2742

and by CRC P ress
2 Park Square, Milton Park, Abingdon, Oxon, OX14 4RN

© 2022 Ian Long
First edition published by CRC Press 2022
CRC Press is an imprint of Taylor & Francis Group, LLC

Reasonable efforts have been made to publish reliable data and information, but the author and publisher cannot assume responsibility for the validity of all materials or the consequences of their use. The authors and publishers have attempted to trace the copyright holders of all material reproduced in this publication and apologize to copyright holders if permission to publish in this form has not been obtained. If any copyright material has not been acknowledged please write and let us know so we may rectify in any future reprint.

Except as permitted under U.S. Copyright Law, no part of this book may be reprinted, reproduced, transmitted, or utilized in any form by any electronic, mechanical, or other means, now known or hereafter invented, including photocopying, microfilming, and recording, or in any information storage or retrieval system, without written permission from the publishers.

For permission to photocopy or use material electronically from this work, access www.copyright.com or contact the Copyright Clearance Center, Inc. (CCC), 222 Rosewood Drive, Danvers, MA 01923, 978-750-8400. For works that are not available on CCC please contact mpkbookspermissions@tandf.co.uk

Trademark notice: Product or corporate names may be trademarks or registered trademarks and are used only for identification and explanation without intent to infringe.

ISBN: 9780367754365 (hbk)
ISBN: 9781032020556 (pbk)
ISBN: 9781003181620 (ebk)

DOI: 10.1201/9781003181620

Typeset in Garamond
by Deanta Global Publishing Services, Chennai, India

Cover and illustrations created by Nadia Long.

Contents

Foreword

After spending almost 50 years wandering through the many universes of quality, safety, risk management, and resilience (all in the healthcare sector), it is indeed refreshing to come across another contribution to the domain of system safety that takes us back to the truly important questions that make a difference. Ian Long's *Essentials of Safety* does that in spades.

You might think that an author writing a book about safety, who takes as a fundamental premise 'This book is not about safety', is starting off on the wrong foot. You would be wrong. I confess that I was hooked from the moment the author challenged the reader to 'understand their why', as the first and most important essential element for those working in the safety realm who are wanting to contribute to creating safer conditions for both workers and consumers (of whatever product is being produced). The question of 'Why' is fundamental and is used by Long as a way of re-directing our attention to the principles that guide us not just in work but also in our everyday interactions with others. Of course, there is no 'correct' answer to the question of why any one of us is engaged in safety 'work' and the answers may often be surprising.

Whenever I teach the SPHERE workshop (which focuses on systemic non-linear analytic approaches to adverse event reviews), I ask participants the 'why' question before talking about any specifics. 'Why are you doing this work?' The answers may be as prosaic as 'the hospital was outsourcing the CSR work and I saw this job opportunity in the quality department' or as inspiring as 'when uncle George woke up after his gall bladder surgery and was missing a leg I just knew I had to do something about this so that others would not have the same experience'. And quite frequently I would be told 'I am doing this work so that I can learn'. That is when I know that I have struck gold. All of the answers are valid and make sense to the person who is responding. Long very eloquently makes the point that understanding our 'Why' is at the heart of effective safety efforts – indeed it is something that can very profitably be incorporated into our professional practice on a regular and iterative basis.

There are many other key points that the author covers in this book. Closely linked to the 'Why' exploration is the concept of adopting a learning mindset in order to become an effective leader and practitioner in the field of safety. Long discusses the importance of such a learning mindset throughout the book and provides examples drawn from his extensive experience in the field. The learning mindset is, of course, a reflection of the inherent curiosity that effective safety practice requires. We are not talking about any kind of curiosity but rather a specific approach to exercising curiosity – sometimes referred to as humble curiosity or curious humility – which leads us to approach any safety-critical situation with the assumption that we know nothing and must start with a clear slate (*tabula rasa*). To properly do justice after harm has occurred requires exactly that, and Long does a good job creating a framework to direct our work.

There are many other key precepts that Long presents – pick up the book and you will be surprised to find them throughout. I must mention one last issue that is extremely important – namely the idea of 'listening generously'. Long proposes this as an essential element for an effective safety practitioner. After more than 30 years in the conflict engagement, mediation, and now coaching spheres, I can attest that this is an important tool and powerful capability for anyone who asks, answers, and understands their 'Why' when it comes to safety efforts.

Of course, the fact that the author refers frequently to the two main innovators and thought leaders in the system safety field (namely Sidney Dekker and Erik Hollnagel) and bases much of his book on their ideas and contributions is another reason to explore this book. Whether talking about ETTO, sensemaking, restorative culture, or organizational resilient performance, you are sure to find much food for thought in this most recent work by Ian Long.

Dr. Robert Robson, MDCM, MSc.,
FRCPC Assistant Clinical Professor,
Department of Family Medicine
Faculty of Health Sciences,
McMaster University
February 9, 2021.

Acknowledgements

It is always a wonderful thing to offer a few words thanking people who have helped on a project. In the case of *Essentials of Safety*, there are a few friends and family that have really helped make this book what it is. My daughter, Nadia Long has done, in my opinion, a brilliant job on the cover and the illustrations. Paige Morrison, also my daughter, has been my grammar fixer-upper. Thanks Nadia and Paige. Kevin O'Kane helped greatly with content, ideas and challenges, especially for chapter 1. The *Essentials of Safety* elements were greatly simplified with many definitions, details and explanations corrected and made easier to understand thanks to James Sullivan and Karen Ross. I had the privilege of working, at various times and places, with James, Karen and Kevin when I worked for BHP. I also thank each and every author listed in the bibliography. They are too numerous to mention but were all significant in the contribution their works have had on my career and my thinking about safety and leadership over many years. A special thanks also to Dr. Rob Robson for such wonderful words in the Preface of *Essentials of Safety*. And of course, Trish, my wonderful wife, who has spent the last few years maintaining the balance between pushing me and encouraging me. She was very supportive as I bounced ideas off her and as I covered our walls in paper and sticky notes during the journey from rough thoughts and ideas into the words of *Essentials of Safety*.

Author

Ian Long has a degree in Science and postgrad studies in Occupational Hygiene. After some 18 years working in various industries and roles, he has worked predominantly in the minerals extraction and processing industry. During the past 25 years, he has worked in oil and gas, chemical industries, and semi-government industries. He has held roles ranging from Safety Advisor to Vice President of Health, Safety and Environment within BHP. He now runs his own company, Raeda Consulting, providing mentoring, coaching, and training in leadership and in incident investigations around the world. He has also facilitated many serious workplace incident investigations. His purpose in life, his 'Why', is to share ideas, concepts, and practicalities in safety and leadership with as many people as will listen, so that people start to think differently and positively about the why, what, and how of the things they do in both leadership and safety. *Essentials of Safety* is his second published book, the first being *Simplicity in Safety Investigations*. This was published by Routledge in 2017.

Introduction

If you have an important point to make, don't try to be subtle or clever. Use a pile driver. Hit the point once. Then come back and hit it again. Then hit it a third time.

Churchill, **Andrew Roberts**

This book is not about safety. In many ways, this book is much more about people and leadership than it is about safety. But then again, safety is all about leadership, so there should not be any surprises there. This book is all about people at all levels authentically behaving as leaders. This is a book designed to prompt thinking. It is not a collection of peer-reviewed scientific papers, nor is it a thesis from a PhD. It is simply an endeavour to explore the common elements that underlie safety with respect to the *individual, leaders* and leadership, the *systems* we have in place, and the workplace *cultures* that all come together to drive safe work – to produce Created Safety as compared to Zero Harm, or safety outcomes-based concepts such as 'no injuries today'. If you are a frontline supervisor, manager, or safety professional, then *Essentials of Safety* is designed for you. It is in some ways a handbook for creating safe work, but it is more than that. It is a book to guide your approach, thinking, contemplation, and then activities around the concept and application of safety.

Before we can get into the common elements that hold safety together – the essentials of safety – I believe we need to start with the word 'safety' itself.

What Is 'Safety'?

I ask this question at all of the leadership coaching and workplace incident investigation training workshops that I run and am constantly amazed at the responses I get. It is not as though safety is a new concept. We have been doing 'safety' for many hundreds of years, and I am sure every manager working today has had safety on their minds in some way, shape, or form for much of their working lives. Having said this, when I ask the question about what safety is, I am intrigued by the fact that most people do not have a definition at hand or on the tips of their tongues. The participants in my workshops tend to think about the question for a while and then come up with a variety of answers that range from things like 'it is all about going home without injuries' and 'zero harm' to 'it is all about risk' and 'it is a mindset'.

There are a couple of definitions that greatly influence how I think about the work that I do, both on the coaching front and in the investigation side of my work. These definitions of safety are from Sidney Dekker. The first is certainly what I cut my teeth on as a safety person. He describes this one as 'Prehistoric Safety', and it goes like this:

'Safety is traditionally defined as a condition where nothing goes wrong, where there are no injuries, accidents, incidents, or perhaps even near misses'.

Zero Harm, TRIF (Total Recordable Injury Frequency), the number of days since an injury, and 'going home safely at the end of each working day' would all fit into this definition. Thinking that aligns with this definition of safety will tend to have us using phrases like install barriers, prevent incidents, put in defences, block this, and stop that. Then I read Dekker's more modern definition for safety. It is:

'Safety is the presence of positive capabilities, capacities, and competencies that make things go right, not the absence of things that go wrong'.

When I first read this definition, my initial thought was 'I have read hundreds of definitions of safety', and so I did not pay it too much attention. Then I took a step back and thought about it some more. As I contemplated its meaning, a significant change formed in my mind. The definition changed my view of safety as it relates to leadership. It changed my approach to 'doing' safety, and it certainly changed my way of approaching and doing workplace incident investigations.

Thinking about this more modern definition of safety drives the front-end-loading of work and safety. It goes to critical controls and what we need to do to make sure things go right today. It is a much more positive mindset to have when doing work. I am not suggesting that barriers to prevent interactions of humans with energy are a bad thing; just that approaching safety with a view to trying to get stuff right is a greater space to play in.

So, to me, 'Safety' is about maximising things going right rather than an absence of things going wrong. It is about people in a system, not a system driving people. And it is about maintaining the balance between thinking and doing.

Driven by this definition, I was keen to explore some questions that I have been preoccupied with over the past few years:

What are the few things in Safety that sit beneath all of the complexity and complicatedness of safety and that we simply must get right?

What are the underlying elements that look through each of the lenses of the *Individual, Leaders* and leadership, the *Systems* we use, and the workplace *Cultures*?

What are the things that tie all of these elements together?

What would it look like if leaders understood and practised these underlying essentials of safety? – What would that workplace look like?

What if leaders of workplaces had the skills, knowledge, and expertise to make sure things go right, rather than having to worry about putting things in place after things go wrong?

What if we could proactively build the fixes to workplace incidents into leaders' thinking, behaviour, and routines *before* we had an incident?

What could all of that look like?

This book is my attempt to answer these and other questions. After much deliberation, reading, playing with others' ideas, talking with peers, going through several boxes of whiteboard markers, and many flip chart pages, I came to the conclusion that it is all about people. That may sound obvious, but it is not as clear as it first seems.

The role that people play in safety is all-pervasive. People *do work*. People *lead* other people who do work. People create *systems* such as procedures, rules, guidelines, and policies that provide guidance and direction for the people who do work. People create *cultures* – through their shared basic assumptions, beliefs, values, ideals, and observed behaviours – that help explain how people do work.

It is all about people. It is about what people think, and it is about what people do as they do their work. It is about people as they lead others and as they create systems, and it is also about how people drive workplace cultures.

All of this is in the cause of creating safe work rather than focussing on the outcomes of that work. The reverse is also true, of course. Cultures that do not drive safe work can also be created by how people think about safety and what they do as they work. Our job here is to explore the thinking and doing that assist in creating a culture, and a set of systems, as well as leadership and individual behaviours that drive or create safe work.

We will be looking at the essential elements that need to be in place to create safe work through the lenses of the *Individual, Leaders* and leadership, the *Systems* we use, and the workplace *Cultures*. I am not going to nor am I even tempted to think about a title for this state of affairs as 'Zero Harm'. That term is way too focussed on the outcome, and I want to focus on inputs – inputs in the sense that people think things and do things. The outcome of that is a workplace where the cultures and systems, including the leaders and individuals, are all driving towards a state of everything going as planned and safe work being achieved. Instead of the outcome-related Zero Harm, we need to talk about the input-related Created Safety. Will we have incidents and accidents? Of course we will. It is about maximising the likelihood of things going right that I want to play with. This is what I want to explore over the following pages.

I am not for a moment suggesting that we should throw the baby out with the bath water here. There are some stunning safety systems, ideas, and theories that should not be disregarded. You may well have heard of a couple of recent names for some of them. Erik Hollnagel talks about Safety I and Safety II. I have recently read an article about Safety III. I have not yet heard of Safety IV and V, but I am sure they are not far away. There is also Safety Differently and the New View of safety along with many others doing the rounds. Many of these are also good fodder for your thoughts.

The early years of my safety career were greatly influenced by Zero Harm as a goal and driver of purpose. I feel that what I am exploring here is Created Safety, the idea that we create safety as we set work up for success and as we undertake that work. I believe that applying the Essentials of Safety will assist us in driving to that state of Created Safety.

A simple way to think about the impact people can have on other people is to explore and contemplate what people are *Thinking* and what people are *Doing*, regardless of whether it is about the *Individual, Leaders* and leadership, the *Systems* they create, or the workplace *Cultures* that are created by them.

There are a small number of elements that keep coming up in conversations around safety, during serious incident investigations, and in the books and papers I have read. These elements represent those things that simply must be in place in order to create safe work. These are the things that we simply must get right.

I believe these essential elements for creating safe work are manifest as 12 essentials of safety, hence the title of the book.

The essential elements split, very happily, into two sets of six: one set under *Thinking*, and one set under *Doing*. I believe it is about maintaining the balance between these two that will make the difference. It is not about the 12 essential elements being a safety system that you should follow. I am simply attempting to encourage you to think differently and positively about the 'why', what, and how of the things you do in both safety and leadership.

A word or two on safety systems: there are many views and off-the-shelf 'safety systems', approaches, methods, models, and practices that people will suggest represent the best way to combat workplace incidents and create safe workplaces. This book is not intended to be one of those. My belief is that many off-the-shelf safety systems and approaches will not work for you exactly as they are presented to you, especially in the longer term. The same is true for what is written here. I believe you need to first understand your own business, industry, and broader culture. Then you need to think deeply about what safety is and what safety is not, in your particular circumstance. As a friend of mine recently said to me: 'Being a master at safety in your place of work, you need to be a master at your work and have a deep understanding of that work, and how your people fit it, and fit together'. This is so true and should be borne in mind as you read this work. Only then can you develop your approach to safety and the creation of a workplace where safety is created each and every time your people do work. Getting this right is not always easy and usually requires a trade-off between efficiency and thoroughness. Safety is a balance between many things. For example:

safety and cost; learning and blaming (after a workplace incident); directive and non-directive coaching (for safety); safety and production; listening and telling (by leaders interacting with workers, especially out in the field); authentic and autocratic leadership. Getting the balance wrong can be devastating to the business, to the worker's health and safety, and to productivity. Getting it right can result in a great culture driving great leadership, or if you prefer, great leadership driving a great culture.

I hope to entice you to think about what safety is for you, and I invite you to set your mind to thinking about ways of balancing your *Thinking* about safety and your ways of *Doing* safety that will help you as you strive to get work done.

My views about what safety means have significantly changed in the 40-odd years of my working life. When I first started working, at the ripe old age of 16, I was a labourer on a dairy farm. I thought nothing of setting the hand throttle on the tractor, climbing over the seat onto the carry-all, climbing up a very wobbly stack of hay bales, and throwing bits of hay down to the cows. After this, I would clamber back down the now smaller, but still wobbly, stack of hay, climb back onto the tractor, turn it around before hitting a fence, and then doing it all over again. This was simply the way we did the work. It was the way my boss told me to do it and how he showed me to do it. It was simply the way work was done. The word 'safety' was not really in my vocabulary at that time. I thought nothing of the hazard or its associated risks.

My first job as a supervisor of workers, some ten years later and with a science degree under my belt, was in hindsight not a lot better in terms of creating safe work. I remember asking an operator to run a machine with its guard off and just told him to be careful and not to stick his arm anywhere near the rotating gears.

Over the past 25 years, as a safety 'professional' or 'practitioner', I have seen safety swing wildly between a worldview that took a purely personal account-ability approach to work, especially when it went wrong, and a systems approach that many interpreted as a no-blame approach, where some feel that whatever went wrong, it was the system and not the individual that was the problem. I do not think that these either/or, black/white approaches are useful ways to approach safety. If we have a firm view of what safety actually is and 'Why' we do what we do in safety, then a more pragmatic, practical, and useful balance will emerge, or at least that is my wish and hope.

It has been said by many current leaders in the safety world that 'safety' is about getting things right and is not about the absence of things going wrong. If we accept this approach, and I certainly do, then we can start to think differ-ently about safety and what our role within it is.

One of the things I encourage you to do as you read this book is to think about how you think about safety. Spend time as we explore the various *Essentials of Safety* elements, thinking about what the topic means *to you*. Think about what each of the elements could look like in your business or in your leader-ship, what could make it work, what could make it fail, and how it meshes or not with you and your 'Why'. Look at things through different coloured lenses, with different hats on, with different perspectives. Get other peoples' opinions and thoughts. Form an opinion yourself. That is pretty much what I did in the creation of this book. I have pulled from numerous sources – well over 120 books and papers – and spent lots of time thinking and having conversations. I also spent a lot of time exploring common models of safety so that I could have discussions that helped form an opinion about what we need to do. I hope it triggers some thoughts in you also.

The intent is not to create an instruction manual, a set of practices that if followed to the letter will always result in safe work or safety. This book is not about a set of technical skills a safety practitioner needs, nor is it an attempt to usurp or replicate any 'Book of Knowledge' work or university course. It is an attempt to explore what needs to happen in the leadership and individual space in the workplace. It is in some ways a recipe book for safety practitioners, supervisors, managers, and other leaders. Just like a cooking recipe book, it contains the ingredients for the creation of safe work, an idea I like to call Created Safety. It contains guidance on how to combine those ingredients, how to think about them, how to mix them together and blend them, and how to overcome problems and barriers in their implementation.

The intent is to foster thinking and a mindset conducive to doing what are, I believe, the essentials of safety.

When I first started writing this book, I envisaged divisions or chapters looking through the various lenses of the *Individual, Leaders* and leadership, the *Systems* we use, and the workplace *Cultures*. As I progressed, however, I realised that they are so intertwined that there is little value in this approach as it is all about people. People are the glue that ties all of these lenses together (Figure 0.1):

The *Essentials of Safety* are a set of individual characteristics, distinctions, attributes, or traits that permeate the workforce at all levels. As mentioned above, the essentials or 'essential elements' talk to each viewpoint of: the *Individual, Leaders* and leadership, the *Systems* we use, and the workplace *Cultures*. It is evidenced by a state where, driven through strong relationships and trust, everybody:

Thinking
- Understands their 'Why'.
- Chooses and displays their attitude.
- Adopts a growth mindset, including a learning mindset.
- Has a high level of understanding and curiosity about how work is actually done.
- Understands their own and others' expectations.
- Understands the limitations and use of situational awareness.

Doing
- Listens generously.
- Plans work using Risk Intelligence.
- Controls risk.
- Applies a non-directive coaching style to interactions.
- Has a Resilient Performance approach to systems development.
- Adopts an authentic leadership approach when leading others.

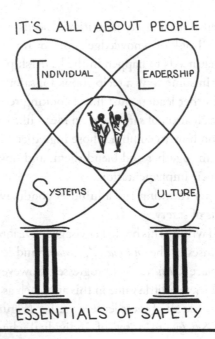

Figure 0.1 The essentials of safety.

Using the above as a framework for thinking and doing (or behaving), individuals, whether they are workers, leaders, technical, or support people (and regardless of their hierarchical level in the organisation), will be internally driven to a state of Created Safety – a set of procedures, systems, behaviours, practices, processes, and routines that align with these elements. It is through the consistent and interrelated application of the 'essential elements' that the workplace cultures will manifest, the systems will be developed, leaders' behaviours will emerge, and individuals will thrive.

It is important not to treat the essential elements as a checklist of things to learn about and do in a particular order. They are best treated as one would a complex system. It is the interrelationships, integration, merging, melding, and intermingling of the 12 elements that will make them work best, all whilst maintaining the balance between thinking and doing.

In alignment with Andrew Robert's quote at the start of this chapter, I will cover each of the *Essentials of Safety* elements a number of times, hopefully from a different angle or perspective each time.

You will see that some of the elements are about internal-to-the-mind activities and some are about behaviours and doing things. It turns out that there are six of each, nearly coincidently by the way. The first six, up to and including

'Understands the limitations and use of Situational Awareness', are all about *thinking*, and the remaining six are all about *doing*. It is a balance.

The *Essentials of Safety* is designed to help you look at things from different perspectives in your drive to create safety in the workplace. And from that perspective I encourage you to think differently about what you do, how you do it, and, of course, why you do it. The elements are just a construct – a bunch of words that portray an idea, a theory, of how things appear to be. It is all about language. There is nothing new in any of this, but I do believe that by distilling and recombining the many views of many people whose conversations and books touch on safety into a simple set of 12 essentials of safety, we can make sure we create safe work in all we do.

To give you a flavour and act as a bit of an index, below is a quick overview and essence of each of the essential elements.

Thinking
Understands Their 'Why'

The more we understand our 'Why', the more we are able to be ourselves. The more we are able to be ourselves, the more effective we become as human beings. Feeling, believing, expressing, and engaging in our authentic selves powers our effectiveness. Our 'Why' describes what is driving us – what underlying purpose makes us do what we do.

Chooses and Displays Their Attitude

Once we *get* our 'Why' – the reasons why we do what we do – we are in a better space to choose how we react to the work that is required of us. We often get to select the jobs we do in a broad sense. We usually choose whether we go to university or not, and if we do, we generally choose whether we study sub-nuclear physics, medicine, or the history of cooking in Elizabethan England. When we do work in return for pay and benefits, and in fact, often at university as well, if you happen to be there, we are sometimes given tasks that we do not really want to do, do not see the value in, or otherwise would prefer not to do. When this happens, we can make a choice – we can choose what attitude we'll bring to the task. We can pick a positive attitude or we can pick a negative attitude. The choice we make can result in a huge difference in how we perceive the task at hand and how others perceive us as we undertake the task.

Adopts a Growth Mindset – including a Learning Mindset

When we have a growth mindset, we understand that we can learn, change, adapt, and improve as we journey towards being the best we can be. When we have a fixed mindset, on the other hand, we tend to feel limited – that we are at the highest level we can attain in the world, that there is nothing else we need to learn, and that this is simply the way it is. We want to explore how we can help move people from a fixed mindset to a growth mindset.

Has a High Level of Understanding and Curiosity about How Work Is Actually Done

If we are a supervisor, manager, or leader at any level, we tend to have opinions and views about how work is done in our part of the business. Depending on who we are and what we do at work, we observe work through different lenses. As leaders and safety professionals, we often write down rules, procedures, standard ways of working, safe systems of work etc., and then we believe that this is how the work is then being undertaken. In the real world, however, the way that the work *actually* gets done does not always match how we imagine it is being done. Those doing the work look through the lens of actual work – this is how we do this job – not necessarily through the lens of how the work 'should' be done. Sometimes an individual does something that is different from the way everyone else is doing it and sometimes the way the work is normally done by many in the group does not match how the written method or procedure tells us it needs to be done. Leaders need to be curious about all three: Work-As-Done – how the work of interest is actually being undertaken by the individual; Work-As-Normal – how the work is commonly done across the broader group or business; and Work-As-Written – how we have documented the work method in procedures, work instructions, rules, standards, etc.. In other words, the way the Work-As-Done 'should' be completed.

Understands Their Own and Others' Expectations

We all have expectations of ourselves and we all have expectations of others. Our goal is to explore how expectations can be formed, shared, understood, and translated into behaviours and conversations. As an example, we cover my thoughts and expectations regarding the creation, use of, and management of procedures. Following accurate procedures thoughtfully works for me. We also talk about the exploration of leaders' behaviours as an indicator of expectations.

Understands the Limitations and Use of Situational Awareness

In the workspace, situational awareness is the ability of a human to observe and understand their environment. Situational awareness relates to being aware of our work surroundings and taking actions that take that into account as we work. The trouble is that we, as human beings, are not capable of keeping an eye on everything that is going on around us – it is simply not possible. Deciding what to keep an eye on and how to keep an eye on it is key to using situational awareness here. Aligning mental models before, during, and after an activity is hyper-important when trying to keep a collective eye on what is about to happen, what is happening, and what did happen. Having foresight is not an easy thing to master, but thinking about resilience performance and chronic unease can help people focus on what needs to be focussed on, rather than trying to understand the entire world.

Doing

Listens Generously

Listening is the most important skill a leader can possess – it is an art and a skill that can be learnt and practised. By *leader* I do not mean the hierarchical leader-as-manager, but the more expansive leader-as-influencer. That would mean pretty much anybody. Generous listening is all about paying attention to, being curious about, and otherwise focussing on the person being listened to, rather than having focus on what the person doing the listening wants to hear.

Plans Work Using Risk Intelligence

Many of you will be familiar with the term 'Risk Intelligence.' I am using it here in a broad sense rather than the tight definition of Dylan Evans. In this book, Risk Intelligence as a concept attempts to encapsulate some of the concepts and ideas of the Efficiency-Thoroughness Trade-Off, Risk Intelligence itself, having a suitable wariness for the effectiveness of controls, and chronic unease. We use all of these, along with many more ideas in something that we do all the time, and it is essential to getting things done in the workplace – that is the practice of planning. Planning is one of the foundational elements of getting the creation of safe work right. There is always a balance, however. We cannot get the planning done to such a level that we know 100% of the time how 100% of the tasks are completed with 100% accuracy. We have to strike a trade-off between being efficient and being thorough in all things, including in our planning. We need to do enough but not too much. We can help strike this balance by understanding what is actually going on, being risk intelligent, and having the right level of understanding that things will not always work out the way we plan them to. Being preoccupied with failure also sits well here.

The best summary of this idea is a direct quote from Chapter 1: 'When things are going well, leaders should worry. When things are going not so well, leaders should worry. Leaders need not be obsessed by what could go wrong, they just need to be preoccupied with it.'

Controls Risk

The path to controlling risk lies, on an individual basis, with always preserving options, applying procedures thoughtfully, and monitoring trigger steps and critical steps combined with activities such as telegraphing deliberate action. These can all make a real difference as we strive to make things go right as we do work. Each of these concepts is explored, along with the recognition that controlling risk is a balance of people trying to keep in mind; why they are doing what they're doing; what the level of situational awareness is; what their level of risk awareness is; what their mental models are; what they put into their planning activities; the risk control measures they choose (hierarchy of control, for example); their expectations regarding failures; preserving options; being mindful; and of course what tools and equipment, procedures, and systems they need to use.

Applies a Non-Directive Coaching Style to Interactions

Coaching and using a coaching style in leadership and management is a wonderful way of making a profound and positive difference in helping people bring out the best in themselves. We explore a commonly used coaching approach utilising the GROW model. We will talk about how it can be used in a pure coaching environment, as well as in the leadership and management space. I have also included what I have found to be a great set of question topics that I review each and every time I prepare for a coaching session.

Has a Resilient Performance Approach to Systems Development

In order to build sound, effective, useable, and accurate systems, it always pays to be logical about it. One effective way of doing that is to apply the framework, or lens, of resilience as you create your systems. In alignment with a strong focus of this book, I want to convey resilience in its positive light, rather than its more traditional *watch out for what could go wrong* approach. Therefore, here we tweak the four elements of Resilience Engineering from being about the negative, or neutral, to being about the positive and go from there:

We change *Respond* from 'Knowing what to do when trouble goes down, or is about to go down', to 'Knowing what to do when things start moving away from going right'.

We change *Monitor* from 'Knowing what to look for or being able to monitor things that could go wrong' to 'Knowing what to look for or being able to monitor things that need to be in place to ensure things go right'. There is no change for *Learn*: 'Knowing what has happened and being able to learn from the experience'. And there is also no change with *Anticipate*: 'Knowing what to expect or being able to anticipate developments into the future'.

We also discuss the idea of a Safety Oscillation model that explores the various elements of a system and aids in helping us decide which links between elements we can strengthen or reinforce and which links between elements we should reduce our emphasis on. For example, we may wish to de-emphasise/disconnect the negative impact an increase in production pressure can have on safety focus, especially as we create or modify our systems.

Adopts an Authentic Leadership Approach When Leading Others

Truly effective, caring, powerful, and sturdy leadership is all about people being themselves only more so. Leadership is about setting context, direction, purpose, and the 'Why' of work. Leadership is all about people being authentic with their stories, their backgrounds, their foibles, and their failures. Intent-based authentic leadership and a strong motivation to help their peers and teams be the best they can be is most assuredly the way to go.

Before we get into the nitty-gritty of the essential elements, I offer a final reminder that the elements simply represent my views of what I think are the essential elements of safety. I believe that they are important to get right, and will make a difference. Is it complete? Probably not. Is it definitive? Absolutely not. Will it drive the reader to think differently about how they think about safety and leadership? I certainly hope so.

Chapter Overviews

Chapter 1 – The Essential of Safety Elements

In this chapter, we explain each of the essentials of safety in detail. I have included examples where this is logical and also key takeaway messages for future contemplation. The intent is to provide a rich abundance of information on each essential element from slightly different perspectives. The main point to remember here is that we are trying to understand the key elements that must be in place as we create safe work. These essential elements talk through the lenses of the *Individual*, *Leaders* and leadership, the *Systems* we use, and the workplace *Cultures*. But at the end of the day, it is all about people and their relationships with things. People *do work*. People *lead* other people who do work. People create *systems* such as procedures, rules, guidelines, and policies that provide guidance and direction for the people who do work. People create *cultures* – through their shared basic assumptions, beliefs, values, ideals, and observed behaviours – that help explain how people do work.

Chapter 2 – Leaders' Perspectives – Practices and Routines

In this chapter, we start to look at what some of the activities a leader at any level of the organisation can practice in order to start to use the *Essentials of Safety* elements in their daily lives. This chapter is where the rubber hits the road. It will provide you with some things to think about and do that will help you develop the skill sets required and demonstrate the advantages of the elements, especially when they are practised together.

Chapter 3 – Barriers and Their Remedies

The *Essentials of Safety* elements are all well and good in theory, but as we know when we come across a different way of thinking about safety, there are always a number of questions that we need to answer. For example: Do they work in practice? What will get in the way of making them effective? What can we do to encourage the elements to be accepted and sustained? For each of the 12 elements, we describe a bunch of common barriers to effective implementation. For each of them, we also provide some thought provokers and solutions, or remedies, to the described barriers.

Chapter 4 – The Essentials of Safety as a Driver of Learning

This chapter builds on the idea that the conversations we have before a workplace incident should be the same as the conversations we have after a workplace incident. We should be interested in how work is being done and we should strive to understand why it is being done that way. I describe, for each element, thought provokers aimed at stimulating your mind to consider which questions can be asked and which themes could be explored to aid learning – both before and after workplace incidents.

I have then, by way of example, expanded on the thinking and ideas from my last book *Simplicity in Safety Investigations* to include using the *Essentials of Safety* elements as a part of an investigation process. I have tweaked the process to focus more on the idea that after a workplace incident we need to spend time working out what we can learn from it rather than simply determining what went wrong and why. This is the reason why I talk about a 'Learning Study' rather than an 'Investigation'. I believe that the outcome is a learning study process that is more simple and effective than the process described in *Simplicity in Safety Investigations*.

Chapter 5 – Assessing and Measuring Success

There are many approaches to assessing and measuring things. Some measurement techniques are quantitative, some are qualitative, and some are a combination of both. In Chapter 5, we attempt to compile some assessment and measurement options for each of the *Essentials of Safety* elements. In turn, we explore methods and ideas for how to measure the level of efficient and effective application of the element in question.

Conclusion

You will have guessed by now that *Essentials of Safety* is all about leadership and the things people do and think about. It is not a pure how-to-do 'Safety' piece of work. This is absolutely intentional. Safety is all about leadership, and leadership is all about relationship. And as Rosa Antonia Carillo talks about in *The Relationship Factor in Safety Leadership*, 'relationship is defined as a biological, mental and emotional connection shared by humans for the purpose of mutual support in adaptation and ultimately survival'. To me, in *Essentials of Safety*, relationship also includes the relationship people have with thoughts and notions as well as the relationship people have with coaching or with authentic leadership for example.

Although I have said in this introduction that you need to ponder what is written here and see if it makes sense to you and helps you think and do safety, in many ways *Essentials of Safety* is a recipe book. And just like a cooking recipe book, it is not the be-all and end-all of the way to do things. It does provide some options on how a supervisor, manager, and safety professional can think about and then do safety. It represents ways of thinking and doing that I believe will enhance the likelihood of you creating safe work in whatever endeavour you entertain. Again, it is about the essentials of safety that underpin or support the 'safety' work done in the spaces of the *Individual*, *Leaders* and leadership, the *Systems* in which we operate, and the workplace *Cultures*.

Chapter 1

The Essentials of Safety Elements

I always learn something when I give a seminar, even if the audience don't

Smashing Physics, Jon Butterworth

The same is true when I was preparing my thoughts around each of the *Essentials of Safety* elements that you see here. I learnt so much as I read, summarised, and then thought deeply about each and every one of the elements briefly discussed in the introduction. It is through getting the balance between thinking and doing right, even when writing a book that deepens our understanding of the topic of interest.

Now that we have had a quick look at each of what I believe are the essential elements of safety, let's now delve more deeply into each one. At the end of each essential element section, I provide what I feel are the key takeaways and where relevant, an example or two. At the end of the chapter is a summary of those authors and the book references from the bibliography at the end of the book that prompted my thoughts and ideas into something that I hope is useful for you, the reader. I recommend these books and authors to you.

Thinking

Understanding Their 'Why'

Knowing our 'Why' drives everything else we do. People who know why they do what they do tend to operate more effectively than those who simply know

DOI: 10.1201/9781003181620-3

1

'what' they do and 'how' they do it. Those who know 'why' they do what they do are able to live by their own guiding principles. Whatever job they have, they don't consider their work 'work' – they consider work as a manifestation of their choices and as a display of who they are and why they do what they do.

Although related, knowing our 'Why' is significantly deeper than having a purpose, set of goals, targets, wishes, or hopes in life. It is a foundational driver or *raison d'être*, if you will, of what we do. It is our reason for being.

The moment we understand our 'Why' is like the moment we turn on a light in a darkened wine cellar: as we flick the switch, the path we need to take is illuminated. Things become clear. And somewhat more importantly, the reason why we are heading in that direction appears before us. In the case of the wine cellar, the path to the fine bottle of Australian Shiraz is apparent. In the broader context, knowing our 'Why' gives us the path ahead, the direction we need to go in, and forces us to think about the 'What' and 'How' of our activities. This is the power of understanding our 'Why' and why it is worthy of our exploration.

To find our 'Why' is not always an easy task. We cannot just make something up on the spot. A person's 'Why' is something that needs to be worked out over time. It needs to be drafted, reviewed, refined, and rehashed a number of times before we get it right. A person's 'Why' is not something that you can simply tell them. If we are their leader, we cannot tell a team member what their 'Why' is and then make sure it is aligned with ours. We need time to allow them to work it out for themselves, just as much as we need to give ourselves time to work out our own 'Why'. Allowing people space and time to work out their 'Why' is critical if we want it to be real for them. Helping understand your 'Why' or helping someone else get theirs is a bit like baking a loaf of bread. You collect the ingredients, knead it around for a while, let it sit and prove, mix it all up again, let it sit around for a bit longer, and finally pull it all together into a fine fresh loaf of understanding, meaning, and direction.

In a later section, I will talk about the need to understand our true self as a part of uncovering what authentic leadership is all about. So it is with the exploration of our 'Why'. In both the exploration of authentic leadership and in the exploration of our 'Why', we need to find our authentic and true self. This is because both are all about a deep and absolute characteristic of self. It does not matter whether you are a C-suite executive, a frontline operator, or a professional in your field, you need to authentically know who you are, what drives you, where you have come from, and where you want to go – you need to get to your 'Why'.

Once you know why you do what you do – or in other words, once you know your 'Why' – it is important to keep your 'Why' alive and up to date. I have refined and tweaked my 'Why' a number of times over the past few years. Talking about your own 'Why' with others and asking others about their 'Why' is a great practice – as long as you don't overdo it.

In a more general way, expanding the concept of 'Why' to other areas of a business also helps to keep the concept alive. I try to always spend some time at the start of significant training and learning sessions to explore the 'Why' of the training. For example, at the start of training sessions on coaching or field leadership conversations, I talk through 'why' using a coaching style in the participants' leadership activities that could help them be more effective as a leader. I also talk about why having effective field leadership conversations can actually make their job much easier and their leadership more impactful. Exploring the 'Why' of the process being learnt can be a great way of enrolling people into the concepts you are trying to help them learn. After all, if the 'Why' of the learning session has some level of alignment with the individual's 'Why', there is a much higher likelihood of learning being achieved.

Another example is when I talk about why we do investigations after workplace incidents. Historically, we were taught that we do workplace incident investigations in order to reduce the likelihood or eliminate the possibility of a re-occurrence of the incident. These days, I do not believe this. Not having a re-occurrence is a side benefit of the investigation process, a bonus if you will. The reason why we investigate incidents is to learn. The 'Why' of a workplace incident investigation could be described as: 'To dig deep into the whats, whys and wherefores of the incident so that we can uncover the lessons that need to be learnt across the broader business so that related activities go right into the future'.

When you are in the field interacting with other people, say in some kind of field leadership conversation process, or simply a peer-to-peer conversation, regardless of your leadership level, take the opportunity to ask people why they do what they do. It is always an easy and fun conversation. When pushed, they often go way beyond the 'for the money' response and talk about the impact that work has on their families and on their lives. With a bit of thinking, people will realise they have a personal purpose in what they do. People who know their 'Why' tend to be able to answer these sorts of questions with more conviction and less deliberation and thinking than others. Also, people who know their 'Why' tend to be more satisfied with what they do because they know why they do what they do and the difference it is making.

In summary, the more we are able to be our true authentic selves, the more effective we are as individual human beings. Finding, expressing, feeling, believing, and engaging our authentic self powers our effectiveness. Being powerfully effective frees up our ability to decide what to do in a situation and this then empowers us to go and do it. There is an art to being authentic. We must understand and be very clear about our differences, what sets us apart, what we contribute, and what works for us and have a strong sense of who we are and where we have come from. This all comes together to establish our unique 'Why'. This is what knowing our 'Why' is all about. This is my 'Why':

'To share ideas, concepts, and practicalities in safety and leadership with as many people as will listen, so that people start to think differently and positively about the why, what, and how of the things they do in both leadership and safety'.

My 'Why' gives me direction and provides a touch stone to test against when I do something new. It gives me feedback when I go off-track and it gives me feedback when I am firmly on track. I can ask myself: 'Is this in alignment with my 'Why' – why I do what I do – or is it something else that may be a distraction to the main game?' I find it supremely beneficial in my life and work.

My 'Why' has helped me and continues to help me greatly in being persistent in writing both *Simplicity in Safety Investigations* and in the three years and multiple drafts of *Essentials of Safety*.

An Example

I was asked to facilitate a session with a small mine as they created their annual Health and Safety plan. The draft workshop agenda included the creation of the draft Health and Safety plan as the 'What' – what we plan to do during the year. And it talked about creating a Charter to describe the 'How' they intend to complete the Health and Safety plan. The Charter was to talk about the ways the team would interact with others, the styles of communication they would use – their rules of engagement as it were. The team had recently had a pretty tough set of culture survey results. They felt a bit lost and were not looking forward to another workshop that led to another set of goals, half of which were likely to end up being unattainable, just like in previous years. I proposed they dedicate the first few hours at the start of the two-day workshop to run a session aimed at discovering the team's 'Why'. I felt that if the team had a common idea of what the team existed for, what contributions they have made to others, and the impact those contributions have had, they would be in a far better place to make sure that whatever they came up with in the Health and Safety plan and the Charter were in alignment with their 'Why'. This is, of course, all classic Simon Sinek stuff. The outcome of the two days was a team 'Why' that read as: 'To build relationships and partnerships that drive cultural change around risk management so that people are empowered to make informed decisions about work'. Beautiful.

KEY TAKEAWAY

Knowing your 'Why' changes what you do and how you do it. It draws the path and lights the way. When you truly know why you do what you do and it aligns with your values and your authentic you, you become unstoppable in your drive, destination, and delivery.

Chooses and Displays Their Attitude

Once we know our 'Why', we have a direction and purpose in what we do and how we do it. We can then start to think about how we might approach the work that we need to do. When we work for someone else, and even when we work for ourselves, we do not always have a choice about the things we do in our jobs. At the end of the day, there are some tasks that we just have to do, even if at first glance we do not want to do them. We do not have to be ecstatic about doing these tasks. But we do need to choose how we react to them. We get to decide how we view those tasks. This is what choosing and displaying our attitude is all about. We get to choose how we react to work.

We tend to know when a task is coming at us. It could be sweeping out the workshop, taking visitors around the workplace, doing the dreaded reconciling of bank accounts, completing a tax return, changing sheets, emptying bed pans, filling in forms, or some other tasks that we have to do as a part of our jobs. For others, it may be a deadline for a report, a meeting with a peer, an assignment due. Whatever it is, one way of handling those tasks that we are not too keen on doing is to spend a bit of time up front choosing your attitude with respect to those tasks. By this I mean to spend a few quiet moments and ask yourself which attitude you will apply to the task at hand. Will you be aggressive, passive, manipulative, determined, rebellious, optimistic, cynical, humble, cautious, or maybe even self-righteous? The choice is yours. Whatever attitude you choose in your little contemplative moment, you can then apply to the task at hand. This can result, especially if you pick a positive attitude, in making a huge difference to how you perceive the task at hand and how others perceive you as you undertake the task. You can make an enormous difference to those around you if you have chosen an attitude that is supportive and positive as compared to an attitude that is negative and disruptive.

Choosing and displaying our attitude applies not only to the tasks we do but also to how we react to other people. We get to choose how we react to those around us all of the time. We do this with friends and lovers as well as with strangers and people we may greatly dislike. When someone says or does something, regardless of whether it is positive or negative, we can react in any number of ways. If we feel that what they did or said is not in alignment with our views and ethics, we may get upset or take it personally. We might ignore them or maybe we react with anger. If we liked what they said or did, we can react supportively and compassionately, smiling and engaging them in further conversation. How we react is completely up to us. We need to remember that how we react is a choice *we* make.

Our attitude to a given situation can be greatly influenced by our 'Why'. If we are clear about why we do something, then the little bits and bobs that go with that activity can be thought of as just something that we do to get to our bigger story – our 'Why'. We are doing the things we do not like in service of getting to make a difference, or whatever your 'Why' happens to be. This approach can make life and work a hell of a lot easier and more fun than otherwise it may be.

Choosing an attitude that represents an understanding of what needs to be done, why it needs to be done, and how it needs to be done is a much happier place to be than choosing a 'whatever' attitude and just doing the task. People who know their 'Why' and live by it seem to be more likely to adopt a positive attitude. People who have a positive attitude know who they are, what they want, and how to get where they want to be. Their behaviours are aligned closely with and are driven by their 'Why'.

When our behaviour is out of alignment with our intended attitude, values, or our 'Why', cognitive dissonance can set in. Cognitive dissonance is when we feel discomfort when two or more ways of thinking or feeling contradict each other. Cognitive dissonance is uncomfortable and is best described by thinking through a couple of simple examples: ordering a takeaway fat-filled burger and then talking about healthy eating to the kids later; believing you are environmentally friendly whilst driving a V8 4X4 gas guzzler; or smoking when knowing that it causes cancer. Cognitive dissonance can have a detrimental impact on our own happiness, our workmate's happiness, as well as productivity and the quality of the work. This is mainly driven by the internal conflict showing up on the outside as words and actions.

It is therefore very important to help your people learn about their attitude choices, cognitive dissonance, and the strong interrelationships between attitude and their 'Why'.

An Example

I recently ran a workshop on learning from workplace incidents using a modified Incident Cause Analysis Method (ICAM) combined with a Work-As-Done, Work-As-Normal, and Work-As-Written Timeline and Five Whys process. We were talking about the impact investigations can have on leaders' workloads, including potentially disrupting business-as-usual activities. It was felt by the group that they wanted to avoid leading or facilitating safety incident investigations at all costs. We spent some time talking about why we do investigations – to learn from them and improve the workplace accordingly. We talked about how our mindset and approach to an investigation can actually embrace our leadership and make a difference that results in real change. Viewing investigations through this lens, they agreed (after a while) that being a part of an investigation that leads to a learning outcome would be worthwhile. In the alternative, they felt that simply going through the process and being mindless of the 'Why' of the process could easily be painful and difficult. It is often said that the way we think about investigations is more important than the process we use. When talking with a couple of the participants a month or so later, they reported that they had gone into an investigation with an open and inquisitive mind and got a lot out of it, both as an independent investigation leader and also as a driver of understanding the incident and what we can learn from it.

KEY TAKEAWAY

You don't always get to choose what you do at work. But you always get to choose how you react and what attitude you take to the work you do.

Adopts a Growth Mindset – including a Learning Mindset

If we are clear about why we do what we do and we are approaching it with a positive attitude, the chances are that we will have a growth mindset that aligns with our 'Why'. Having a mindset that aligns with our 'Why' sets us up for success.

A mindset is simply a state of mind, a way of thinking, looking, or reacting to a situation or state. The two I will principally talk about here are a growth mindset and a fixed mindset. We generally talk about people either having a fixed mindset or a growth mindset. A fixed mindset is limiting. A growth mindset is infinitely expanding. We include in the arena of a growth mindset the idea of a learning mindset. Taking it one step further, Simon Sinek talks about an infinite mindset, where we take a much longer and wider view of the world.

Folks with a fixed mindset believe that people, including themselves, can absorb skills only to a defined level (e.g. 'I could never learn to juggle') or that their intelligence is limited up to a certain point (e.g. 'I haven't got the intelligence to be a doctor'). They often have a static view of how to handle complexity or believe they have limited capability to improve because they think they have hit their maximum potential. No matter how much effort and time they put in, they feel they cannot become what others think they are capable of.

People with a growth mindset, on the other hand, understand that stuff can change. They have a belief that people, including themselves, can learn, reshape their intelligence, adapt their approach to complexity, have the capability to improve, and with effort and time can achieve much more than they are currently achieving. Those with a growth mindset recognise that there are always opportunities to learn and improve. These two mindsets are clearly at opposite ends of the pole and it becomes very obvious very quickly which mindset a person has.

The important point to remember is that people's mindsets are not stable in the long term. People can choose to have a fixed mindset or they can choose to have a growth mindset and this can change over time. This choice is what makes all the difference. It is vital that people understand that mindset is a decision and based on belief, attitude, and their 'Why'. After all, mindsets are things that occur within the mind and as we all know, we can change our mind.

It makes sense, therefore, that we should want to have a fair proportion of people in our teams possessing a growth mindset. To get to that point we should not rely on luck but rather on good management. Helping establish a team of people with growth mindsets cannot start on day one of each person's employment – it needs to start during the selection and recruitment process phases. During early screening and interviewing, it is possible to determine whether the prospective employee believes they can learn more and grow and whether they have a growth mindset. To me, this needs to be an integral component of the recruitment process.

The creation of a team with a high percentage of growth mindset thinkers also needs to extend beyond the original employment selection process well into the daily life of each employee. By specifically calling out growth mindset during inductions, regular performance appraisals, task assignments, development planning sessions, etc., you can assess whether the person is up for some challenges, whether they are okay to explore some of their failures, are eager for feedback, and generally whether they are keen to stretch themselves to the next level. These traits should be encouraged and prioritised in conversations around an individual's development.

One manifestation of a growth mindset is associated with learning. We can use the term learning mindset for this. In addition to helping our people develop a growth mindset, it is useful to cultivate a learning mindset throughout the various training and learning processes and activities.

One easy way to do this is to talk about learning during conversations, inductions, activities, workshops, learning studies (incident investigations), and field leadership conversations. This can be achieved by talking about the 'Why' of a training/learning workshop at the start of the session. This really helps set the scope and encourages better learning and engagement in the workshop, especially if the 'Why' of the session strikes a chord with the 'Why' of the participant. I also ask this question at the end of any workshop I run: 'As a result of the conversations we have had over the last hours/days, ask yourself what you will do that is different? or what you will do differently after today?' The intent is for the participants to understand that learning is about changes in behaviour driven through a change in thinking and believing new concepts and ideas. I want them to think about whether they have actually learnt anything.

We need to encourage and promote the idea of thinking about how we learn. That includes learning how to learn better – of rethinking the way we do things, how we operate, and how we learn. After all, as Steven Poole has said: 'If we are not rethinking ideas then we are not really thinking at all'.

We should always encourage people to think and challenge the way we do things and why we do them. This, in many ways, is an element of the GROW model of coaching that I will talk about later in the chapter. The practice of coaching helps people think about how they are doing things and through great coaching questions, coming up with better ways of being better.

I would like to share a couple of ideas that have come from such coaching conversations which really highlight the power of having a growth mindset and desire to learn.

One stems from a coaching conversation between a manager as the coach and a superintendent as the player that was shared with me as I was coaching the manager a few years ago. They were talking about workplace incident investigations and their relationship with day-to-day work. The superintendent suggested that we need to learn from how we work on a day-to-day basis and not just after a workplace safety incident. As a part of the coaching session, the superintendent came up with an option for how this might look. He suggested doing a simple, resource-light investigation after a field leadership conversation that had identified an issue. I later caught up with the superintendent and we chatted about the concept and how it might work. It aligned with some work I had already been doing on the subject and the idea was developed into a process where we can learn from normal work. I call these learning from normal work (LNW) reviews. The idea is that over several field leadership conversation activities, we build up a story about Work-As-Done, Work-As-Normal, and Work-As-Written related to a specific element of a task. We then get a small team together from the work crew and explore the differences between Work-As-Done, Work-As-Normal, and Work-As-Written, and what we can learn from them. This is done before an incident occurs, not after the workplace safety incident has manifested itself. It is all about learning from adjustments that are made day-to-day during Work-As-Normal. The process is covered in detail in Chapter 4.

This sort of activity (LNW review) cannot easily happen if the leader driving it and the team doing it have fixed mindsets. They need to be willing to explore their understanding and be ready to learn together about what could be achieved.

A learning mindset can manifest during safety-related incidents and investigations into them. An organisation that possesses a growth mindset will endeavour to learn from incidents. There can sometimes tend to be an element of wanting to punish people for making mistakes that result in workplace safety-related incidents. This can be minimised or avoided – and the incident can be learnt from – by asking questions differently and having a growth mindset approach to incident investigations generally.

When an incident occurs, the way we ask questions, what we focus on as leaders, and how we react can all have a big impact on the ability of the business to learn from the incident.

Below is a short list of the sorts of questions that I encourage leaders to ask after a workplace incident. They are driven by a learning mindset:

■ Are our people okay?
■ Are things safe and secure?

- Who has been harmed? (first and second victims)
- What are the needs of those harmed?
- What should be done to satisfy those needs?
- What is the narrative of the event? (i.e. tell the story)
- What went well?
- What went according to plan?
- What could have happened (that didn't)?
- What else do I need to know?
- What can we learn from this incident?

In order to focus on learning during an incident investigation, it is important to remember that human performance is variable and that unexpected events come from good as well as bad decisions.

This means that we should not simply focus on bad decisions when investigating incidents. Bad outcomes do not always come from so-called bad decisions. Good outcomes do not always come from good decisions. Human performance variability is neither positive nor negative – it just is. We need to seek to learn, regardless of what opinion we have of the decision made by an individual that may have played a part in the incident. It is important to learn from major incidents. Applying a learning mindset to low-level incidents as well as to major level incidents is a very sensible approach. There is often a lot to learn. The decision to learn from an incident should not be based solely on the level or magnitude of the outcome. It is even better to learn from Work-As-Normal – when things go right as we talked about earlier on LNW (learning from normal work) reviews.

Another element in the drive to encourage learning is to encourage people to 'speak up'. And to support this, perhaps more importantly, leaders should be encouraged to listen. I saw one business change its 'speak up' programme aimed at the working level employees to a 'listen up' programme aimed at leaders and encouraging them to listen and react more when workers speak up. This sent a strong message that the leadership was authentically interested in listening and learning when others have something to say.

This 'listen up' approach is intrinsically related to leaders deferring to expertise. Leaders need to realise that if they are experts in one particular field, it does not mean that they are smarter in all other fields as well. Encouraging leaders to listen to, and learn from, experts helps greatly with relationships, authenticity, and trust.

Inclusion of discussions on fixed and growth mindsets in leadership development activities is also an important component of getting the essentials of safety right. Talking with leaders about how their mindset can impact the safety of their team and whether they have a growth mindset – believing they can all learn, get better, and grow – or whether they have a fixed mindset – where things are simply how they are and that is that – can greatly impact leaders' thoughts and hence their words, their actions, and their behaviours. A growth mindset or a fixed mindset can also manifest in the way leaders search out for and listen to expertise. If we have a fixed mindset and believe that we know all we need to know and do not defer to expertise where it resides, trouble ensues. You should strongly consider only promoting people to leadership roles if they have a growth mindset. Having said that, a fixed mindset can be expanded into a growth mindset with some work and will. Coaching and supporting in this space is well worth the effort. Leaders simply being aware of the two mindsets (fixed and growth) helps them recognise they can change – if they want to.

People with a growth mindset tend to pay close attention to experiences and challenges that will stretch them, furthering a belief that brilliance is not the key to success – a growth mindset and passion is. On the other hand, leaders with fixed mindsets seem not to feel happy or strong when significantly challenged – they can tend to lose interest. They can also tend to take their personal failures as a declaration that they are failures. People with a growth mindset meanwhile tend to view their failures as lessons and opportunities for learning.

For all of these and many other reasons, helping people, especially leaders, to evolve from a fixed mindset to a growth mindset needs to be a critical component of everyone's personal leadership development plan.

Historically, many businesses and industries have been pretty good at sharing information after workplace safety incidents, but not many of them are that good at true learning. By true learning, I mean changing the way people think and then subsequently changing behaviours. A part of this is driven by the business creating too many lessons to be learnt. The other common problem is that the lessons local incident investigation teams come up with are too specifically incident-focussed rather than business-wide. For example, 'When buying new forklifts, make sure you understand the blind spots' as compared to 'When buying new equipment it is important to understand the nuances of the equipment, the guidance of the original equipment manufacturer (OEM) and to evaluate any differences between the existing and the new equipment'. The former is only useful if you have a forklift. The latter applies to just about any business or industry and is far more useful as a lesson.

So, a solid way to apply a learning mindset to a business needs to be focussed initially on exactly what the lesson to be learnt is. Language becomes critical here as we have just seen. The lesson could come from a workplace incident, an audit, a recent change, a part of the business that has had excellent or poor performance, or in fact from many other activities. The approach I find best is as follows: the lesson needs to be worded in a way that it is provocative – something that challenges the status quo or existing mental model of the work or topic and excites leaders into action. For example, 'procedures must be SUE – Simple, Useful and Effective' is not very provocative. 'In the real workplace, Work-As-Normal does not always equal Work-As-Written' is perhaps an improvement, but only just. The wording needs to prompt some inquiry as to Work-As-Normal in the part of the business where you want the learning to occur. As an example of a leader with a learning mindset, if the first response to reading a 'lesson to be learnt' from a General Manager, senior line, or functional HS manager is 'We absolutely need to check this out for our bit of the world', or 'There is no easy answer for this'. Then we are at the right level. Of course, to make sure a lesson to be learnt reaches a state where it does attract attention also relies on a learning mindset amongst those who created it.

Once the lesson has been worded appropriately, we need to work out what the best way is to help the people learn the 'lesson to be learnt'. This may be very different for different parts of the business as well as for different levels in the business and also different for various professions and worker groups. It may also only apply to a select group or team within the business.

We know that it is important to consider who the audience of a learning activity will be and then tailor the learning approach to match that audience. For example, some lessons to be learnt will be better suited to the supervisor level in the organisation because they are the ones who orchestrate work, approve risk assessments and task-based risk assessments, and also provide frontline guidance and leadership for the achievement of safe work. At other times, the lesson will need to be aimed at the broader audience of those actually doing the work.

What we do to impart the lessons needs to be in a language that resonates with the audience, is meaningful in a manner that fits with their worldview and mental models of the work, and results in 'discovery' of the learning rather than a rote learning of it. If they get their own 'why' in relation to the learning, the learning becomes important for them. It is only then that it will result in changed mental models, thinking, and behaviour.

Once you have gathered enough information to form an opinion on the target audience/s, it is then up to the leaders to share the lessons in such a way that promotes learning and the required changes in behaviour. The work of driving the change so that the lesson is learnt needs to explore the current mental model of the work and where it needs to get to. It could explore Work-As-Normal and Work-As-Written in relation to the topic of the lesson and what the current drivers are for Work-As-Normal, especially if it does not align with where the Work-As-Normal needs to be for the lesson to be learnt. We will talk more about Work-As-Done, Work-As-Normal, and Work-As-Written later in the chapter. For now, think about Work-As-Done simply as how work is being done on any specific day, Work-As-Normal as how the task is done by others on a routine day-to-day basis, and Work-As-Written is what the rules, procedures etc., say about how the work 'should' be done. In many ways, considering a lesson as being about changing behaviour is useful as this is what learning is all about. This leads it to be open to an investigation or learning study-style approach, where the teams get together and explore Work-As-Normal and Work-As-Written.

Get a group of say ten people together and after explaining the concepts of Work-As-Done, Work-As-Normal, and Work-As-Written, create sets of Elements of Interest where gaps or differences lie between Work-As-Normal and Work-As-Written as it relates to the lesson to be learnt. Then you can run a simple Five Whys process for each Element of Interest. Don't forget to explore the mental models associated with the current Work-As-Normal, and instead of a list of actions, come up with what routines might look like to promote alignment between the old Work-As-Normal and the new Work-As-Normal. Also consider what barriers may be present that may prevent this from occurring, what changes need to be made to their mental models, beliefs, and systems. This session will probably take an hour, including some formal time for discussion, thinking, and contemplation.

Always remember that learning is difficult when the teams are harried or are rushed. So, build some soak time into the process so that the participants can discuss and self-reflect on the learning and what it means to them specifically. Doing this sort of work shows that a business has a growth mindset and more specifically a learning mindset.

An Example

After a potentially fatal workplace incident involving a confined space and an atmosphere that was not suitable for healthy working, the investigation team came up with the following lesson they felt the rest of the businesses needed to learn. It was worded as follows: 'For gas monitoring in confined spaces to be an effective control, we need to consider the method of monitoring including where to monitor and the frequency of that monitoring'.

Approach to the Learning: The various sites of the business got the certified confined space officers and a few other people who use gas monitors to clear confined spaces together. After explaining the incident, the lesson to be learnt, and the concepts of Work-As-Done, Work-As-Normal, and Work-As-Written to the group (including the idea that learning is about changing ideas, mental models, and behaviours), they explored together how they normally use gas monitors in the field when confined space testing was being undertaken. This became the 'old' Work-As-Normal. They then talked about whether what they actually do (the 'old' Work-As-Normal) matched or did not match the intent of the lesson. The testers came to the realisation that the way they currently worked could result in the same issues spotted by the investigation team. They then worked out how they could approach their work, what was needed to change in their mental models and behaviours, and what the business leaders could do to help them achieve the changes they (the workers) wanted.

KEY TAKEAWAY

Having a growth mindset expands. Having a fixed mindset limits. We can change a fixed mindset to a growth mindset – if we want to.

Has a High Level of Understanding and Curiosity about How Work Is Actually Done

Eric Hollnagel reminds us that as managers, we need to understand that the way work is done out there in the real world is often not exactly the same as the way we imagine it is being done. If you have read *Simplicity in Safety Investigations,* you will have seen how I adapted the idea of Hollnagel's Work-As-Done and Work-As-Imagined to talk about needing to understand how work is done (Work-As-Done), how everyone else does the work (Work-As-Normal), and how the written procedure or other documentation tell us how to do the work (Work-As-Intended) as a precursor to understanding the differences and hence the contributors to an incident. In hindsight, and in response to clients' comments and suggestions, changing Work-As-Intended to Work-As-Written seems to strike a better chord with users of the investigation approach so I will use that terminology in this work. I read a very interesting tweet from Stephen Shorrock recently that talked about the large number of Work-As- ... that are used today. The graphic in the tweet (from humanisticsystems.com) showed many:

- Work-As-Imagined.
- Work-As-Prescribed.
- Work-As-Disclosed.
- Work-As-Analysed.
- Work-As-Observed.
- Work-As-Simulated.
- Work-As-Instructed.
- Work-As-Measured.
- Work-As-Judged.

And, of course, here are the ones I have used:

- Work-As-Intended.
- Work-As-Normal.
- Work-As-Written.

This really shows us that we can talk about work through a wondrous array of lenses. I have simply chosen the three that seem to work for me here. Viewing work through many of these lenses can at times help us understand the workplace from differing perspectives and that can be extremely useful, especially when trying to understand what went wrong in a workplace incident and what we can learn from it. The three that I am focussing on are Work-As-Written, Work-As-Done, and Work-As-Normal.

Work-As-Written (WAW) represents that vast amount of detailed information contained in all of the procedures, work instructions, work orders, rules, commandments, checklists, injury prevention principles, Task-Based Risk Assessments (such as Task Hazard Analyses, Job Step Analyses, Job Safety & Environmental Analyses, etc.), and many other such instructional documents that drive work. Work-As-Written is often a complex and intertwined set of requirements.

Work-As-Normal (WAN) represents how those involved in doing work in the business normally do the tasks-of-interest that we are focussing on in the conversation or learning study. Details of Work-As-Normal are usually obtained through conversations with those who do the task-of-interest on a daily basis. We always use adverbs such as 'normally' or 'it is common practice to ...' when talking about Work-As-Normal.

Work-As-Done (WAD) represents how the work is actually done at the time in question. This could be when a leader is out having a field leadership conversation, or it could be about activities leading up to a workplace incident being investigated. One thing to keep an eye on when exploring Work-As-Done, whether part of an investigation or as a part of day-to-day leadership, is the concept of drift. Practical drift is when the way the work is being done today has changed over time. This occurs frequently, especially when we have people teaching each other how to do the work. People pick up ways of working that may be easier or quicker and over time that way of working becomes the way we work around here (Work-As-Normal). Drift can also appear as procedural drift in Work-As-Written documents where we have tweaked procedures over time due to periodic reviews or changes after workplace incidents. This can sometimes be to the detriment of the level of control we have over the risks. We can end up with a procedure that misses the mark on critical controls and actually adds risk to a task.

Work-As-Written can sometimes get a bit out of hand in terms of its volume, complicatedness, and complexity, and it is worth spending some time on these aspects of the term. I was at a client's operational site late in 2019 and was told that the business had over 20,000 documents in their system. Sorry, but this is absolutely crazy and unworkable. No human is capable of getting and keeping their heads around such a high number of system documents, procedures, etc. Work-As-Written can get complex and it can get complicated. We recognise that complexity, or more specifically an increase in the level of complexity, through changes, planned or otherwise, can encourage the creation of more complexity and increased risk of things going wrong – not as planned. This will often manifest as a misalignment between Work-As-Done and Work-As-Written. Complexity in systems and procedures can leave end-users of those systems bewildered as to what to do. This can result in both Work-As-Done and Work-As-Normal being different from and inconsistent with Work-As-Written.

Complexity and safety are beyond adversaries and are causatively related. Complexity is inversely proportional to safety. I am not sure if some inverse square law applies here, but certainly when complexity goes up, safety has a propensity to go down.

A quick reminder on complex versus complicated. Rocket science is simple when compared to getting safety right. Rocket science is complicated and difficult to master, but not complex. In a complex modern workplace, it is not possible to create a system that will cover all workplace scenarios. It is, on the other hand, possible to build a system that will get a man to the moon and return him safely to the Earth. The science is tough, especially orbital mechanics, but it is defined and structured. This is why we constantly seek to build resilience practices and adaptive skills in our people as these are essential for humans to operate in a complex environment.

One remedy for both complexity and complicatedness is simplicity and the simplification of Work-As-Written. We need to always look at work simplification with a complexity lens. Understanding that simplifying an operation can often improve reliability as well as safety is important, but it's also important to understand that simplifying too much is not good either. Extreme simplification is not the way to go. Having said that, on balance we should have a reluctance to simplify. And then, when we do simplify, we need to do it very carefully and consciously.

We need to be especially tuned into any changes to systems that are undertaken after a workplace incident and its subsequent learning study. This is especially true if the incident 'cause' was related to the interrelationships of systems or sub-systems rather than some broken widget-type of incident. We need to watch out for putting extra barriers in place after an incident as this does not always do a very good job after incidents that were driven by system interrelationships. After all, it is important to remember that removing a component or barrier out of a complex system does not cause it to fail and adding an extra defence or barrier to a complex system does not mean it will now not fail.

Introducing improvements and changes to systems and procedures can create new ways in which people need new skills, new routines, and new activities. This can introduce new ways in which they can fail, especially until such time as the new skill, routine, or activity becomes embedded in their approaches and the way they do work. We should actively search out and destroy any 'improvements' that add to the already over-burdened bureaucratic systems and introduce more clutter and unnecessary system-driven requirements. We must constantly keep an eye on Work-As-Written.

Checklists as Work-As-Written also come to the fore when we are faced with complexity in the workplace. They provide an aid for judgment; a procedure to follow when the memory of the expert actor is not sufficient to handle the complexity of a task; a verification that stuff gets done in the order it needs to get done in; a set of reminders for the critical bits of the task; a buttress to the skills of the expert.

Checklists are limited and only useful for specific tasks to help experts get it right each and every time. It is important to make sure they are used sparingly and accurately. Keep them short, aiming for a maximum length of ten items.

Overall, we should also talk about the ideas of Work-As-Done, Work-As-Normal, and Work-As-Written during field leadership conversations, when talking about production and when talking about workplace incidents.

A simple way to explore the ideas behind Work-As-Done, Work-As-Normal, and Work-As-Written with people who have not seen or heard of them before is to talk through our own experiences as examples. I am making an assumption that you have all seen the following traffic control device (Figure 1.1).

In all the countries I have worked in, there has been a piece of legislation that tells us when the little person is red not to cross the road and when the little person is green to cross the road but endeavour to not get run over. The language varies and is usually translated into legalese, but the intent is pretty much the same. So, in this case Work-As-Written is the piece of legislation that pertains to the use of the pedestrian crossing lights.

I live in the suburb of Subiaco, which is in Perth, Australia. There is a set of pedestrian traffic control devices similar to that shown above about 200 metres from my house. Imagine that I am standing at the intersection on a cold, rainy, Sunday morning at 0700 and am watching the people as they cross the road. Am I likely to see 100% of the pedestrians crossing the road displaying 100% compliance to the procedure for crossing roads? In this case, the pedestrian lights legislation? Ah, no. I don't think so. I am going to see a variety of behaviours ranging from those looking at their phones the whole time and just walking when they hear the Beep Beep Beep. Others will dodge traffic as they cross regardless of the colour of the lights. Some will stand there patiently until the little person goes green and then look both ways before walking and others will run if there are no cars when the little dude is red or is flashing. This is

Figure 1.1 Pedestrian traffic lights.

Work-As-Normal. The way I crossed the intersection as I went to set up with my little clipboard, and started watching and counting people is what would be called Work-As-Done.

Another way to think about Work-As-Done, Work-As-Normal, and Work-As-Written is to talk through Figure 1.2.

The underlying hazard is just that – a hazard that is presenting various levels of risk over time. Sometimes the risk is lower and sometimes the risk is higher. Look next at the solid horizontal line marked as Work-As-Written. This is depicted as a straight line and represents the way that management has decreed that the work will be done. It could be a work instruction, procedure, set of rules, standard, Job Safety Analysis, work order, Task Hazard Analysis, Standard Work Instruction, etc. Basically anything that tells you how to do a piece of work. You will notice that at no time does the Work-As-Written line come in contact with the Hazard. This sends the message 'Follow the rules and you will be safe' that we often hear. You just need to look at some of the significant incidents in the past to know that following the rules blindly can be a very dangerous thing to do. I recall a few times where I have almost begun crossing pedestrian lights when the little person was green only for a driver to cross the red traffic light in front of me. Even those who simply walk when the little person is green are taking a considerable risk.

We know of course that not all work in our workplaces is undertaken in a manner that exactly equals the way it is written (Work-As-Written). Sometimes it is done more safely, more productively, with higher quality perhaps, and sometimes it is done less safely, less productively, and with lower quality. The vast majority of the time the end result is that nothing goes wrong. This is what we

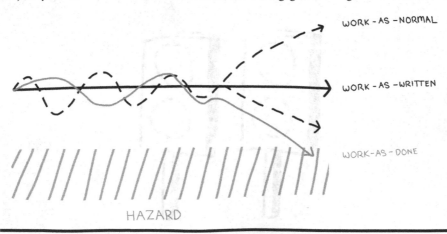

Figure 1.2   Work-As-Done, Work-As-Normal, and Work-As-Written.

call Work-As-Normal and is represented by the dashed lines. We get it right nearly all the time. In fact, whenever you go and watch people do real work, you will see a variety of behaviours; some following the rules precisely, others deviating slightly (for all sorts of good reasons), and still others deviating considerably from the Work-As-Written. Our people who do work day-in and day-out, those who face the various hazards of the workplace are nearly always controlling the risks without getting hurt. They do it not because they blindly follow the rules but because they adapt to the situations as they arise and use their skills, experiences, and knowledge to get the work done. When we look at the third wavy line, we see that the Work-As-Done line drifts away from the Work-As-Normal and Work-As-Written lines towards the hazard. This represents an incident or a near miss. This is how the work was actually done at the time of the event. It is what we would conventionally call a timeline of the incident. It could also represent the way the work was being done as we observed work during a field leadership conversation. At the end of the day, both before and after an incident, we need to be very interested and curious about how the work is being done (Work-As-Done), how others normally do the work (Work-As-Normal), and how we may think the work is done as described in our documentation (Work-As Written).

An Example

Although I started using Work-As-Done, Work-As-Normal, and Work-As-Written to help an investigation team develop a timeline and then focus on the important elements of an incident back in 2014, within a couple of years I was approached by a senior leader of a large client who mentioned that other leaders in his business were starting to use the ideas of Work-As-Done, Work-As-Normal, and Work-As-Written in their normal business conversations. They had started to formalise it during their field leadership conversations and found the ideas to be great and easy conversation starters. They could have conversations about how the task they were chatting about was being done, how other workers and their peers normally did the work, and what their thoughts were on the procedures that make up Work-As-Written. They found this a very powerful tool to help them be effective as leaders.

KEY TAKEAWAY

To get better at 'safety' you need to deeply understand how real work is undertaken, rather than just how you think it is being undertaken.

Understands Their Own and Others' Expectations

Whatever occupation we have, wherever we sit within the hierarchy of an organisational structure, however we interact with people, we all have expectations. We have self-imposed expectations regarding our own behaviours. Other people also have expectations regarding our behaviour. We have expectations regarding how we are treated in the workplace. Other people have expectations regarding how we treat others. And we also have expectations regarding how we treat others, including those who report to us in the workplace, as well as our bosses.

It is a critical component of creating safe work that we understand both our own and others' expectations.

Keeping those expectations to ourselves helps nobody. It helps neither us nor others in the workplace. If others know our expectations and we know theirs, then we are all in a better position to understand what is going on around us – what other people are doing and hopefully what the shared mental models of the work are.

Setting and sharing expectations is a critical activity that never stops, especially for a leader. The simplest way of sharing expectations is by talking about them. A leadership team needs to have open and authentic conversations within the team about the expectations they have of each other.

Another great time to explore mutual expectations with a direct report is during the regular performance appraisals that leaders undertake with their people.

A big driver of the behaviours we exhibit as we share our expectations has to do with our brains and the way thinking can further influence our thinking. The more frequently a thought or a pattern of mental activity occurs in our minds, the more entrenched the neural connections become. With enough repetition, this can become permanent. It is as if the pathways in our brains are like paths in the bush or in the forest; the more people that hike along them, the more the paths become entrenched as paths. The more these paths are used, the easier they become to follow. So the more we verbalise our expectations, the more we think about them and the more aligned our behaviour is with our verbalised expectations. They become very consistent over time as a result. This is driven by neuroplasticity – the ability of the brain to change as a result of changes in what we think, say and do. This is why, through routines and thinking, 'behaving like a great leader until you become a great leader' can be an effective approach. This is not to suggest faking it is the way to go, but rather that we can change who we are by thinking and behaving differently over time and with effort.

One area that I believe leaders need to be crystal clear about is in the displaying and discussing of their expectations in the province of 'just culture'. We need leaders at all levels to recognise that incidents come from normal people doing normal work and not from individual failures, violations, poor choices, or loss of situational awareness. If you have or are contemplating setting up a 'just culture' process to explore where culpability lies after an incident, you need to keep this in mind. It is often not the structure of the 'just culture' decision tree that is the issue that drives frustration and an 'unjust culture' in this space. It is the inconsistent application of the 'just culture' process by the various managers and leaders who apply it that can cause problems. It is by talking through our expectations about how people are treated when something goes wrong – consistently and openly – that a culture that is just can be created.

An interesting bit of brain science helps us understand the drivers to blaming versus learning. Due to neuroplasticity, if you are used to finding people to blame after an incident, you will find your attention easily drawn towards facts and information that point to someone at fault. If, however, you are used to finding the learning out of an incident, you will find information that supports that. This is what we strive for. This is why I suggest shutting down blame conversations and encouraging learning conversations. Ask 'What can we learn?' instead of 'Who can we blame?' At the end of the day, you cannot 'apply' a 'just culture' – you either have a culture that is just or you don't.

Having expectations around how we react to a team member being involved in an incident is also very important when we think about leadership and the impact we have on others. An example is having a supervisor who insists that the person involved in an incident is accident-prone, is a bad apple, a maverick, or they are the reason for the incident. When this happens, an approach that seems to work well is getting the supervisor to rhetorically, and maybe also literally, answer a few questions:

- Who hired the maverick?
- Who trained them, and how?
- Who has supervised them, and how?
- Why has it taken an incident to raise the issue?
- Who allowed them to stay in the business, and why?
- How were they actually trying to make things go right at the time?
- How can we learn from answering these questions, rather than blaming and therefore not learning?
- What can we do to rebuild trust in this individual?

When talking about expectations, whether they are our own or someone else's, we need to make sure that our communications are affective; i.e. they are sent *and* received as intended.

A way of sharing expectations effectively is through displays of behaviour. It has been said by Edgar Schein and others that leadership and culture are two sides of the same coin. This manifests here in the behaviour of leaders at all levels. The behaviour of leaders will create the cultures we strive for, and the behaviours of leaders will be influenced by the cultures we have. Sharing expectations through displays of behaviour can be far more powerful than just talking about them. It moves from 'Do what I say, not what I do' to 'Do what I do'.

This is limited to no value in trying to force the creation of a culture. It is far more powerful to let leaders' behaviours – what they focus on and how they react to things – tell the story about who you are as a business and what you stand for – your culture.

It is recognised that senior leaders' behaviours have a much greater influence on the culture of the place than do the behaviours of middle managers and frontline supervision leaders, who to a greater degree tend to live within the culture rather than create it. So it is very important to spend time focussing on how we behave, rather than telling others how to behave. We need to constantly share the expectations we have of ourselves in this regard.

I encourage leadership teams to annually undertake a somewhat self-reflective internal review and check out the behaviours they exhibit. This will also help call attention to the expectations that the leaders display relating to things in the following categories:

- Visible leadership.
- The level of strategic importance that safety has in the business.
- The quality of our procedures and work instructions.
- Application of the Management of Change process.
- Interest in close-out of actions from safety-related incidents.
- Compliance with local regulations.
- Clarity of roles and responsibilities.
- Measurement of safety inputs (as compared to outcome measures such as Total Recordable Injury Frequency, or the number of days since an injury).
- Handling of conflict.
- Leadership development.
- Leaders' mindsets concerning safety (growth or fixed).
- Attitude to mistakes.

- Whether people see safety as an external, technical requirement, or an aspect of what they do to succeed.
- Leaders' views of procedural compliance and operational discipline.
- 'Blame' or 'learn'.
- Leaders' reactions to failure.
- Resilience as a way of being (culture of resilience).
- Effectiveness of the 'Safety' department – coaches or policemen.

The idea of the reviews is to explore the behaviours that drive our cultures. Getting these behaviours to be the best they can be will ultimately drive the cultures we strive to display and the associated expectations of ourselves and of others.

An Example

I feel that it is very important to be specific in communicating our expectations. An example is in relation to people's expectations with respect to procedures. What follows is an approximation of how I view and talk about my expectations surrounding 'procedures', what I expect with respect to any 'drift' people experience, and my thoughts on compliance.

My View of Procedures

Procedures are very interesting tools of the trade. On one hand, they play a critical role in the creation of safe work. On the other hand, they can be the bane of our existence. There are usually too many of them; they can't always be followed; they can't easily handle unexpected interruptions; they can't guarantee safety; and they are usually not accurate. Sometimes they are simply not very smart – like requiring safety helmets to be worn between open car parks and offices. They are often written by people who do not do the work.

I clearly understand and talk about the fact that no set of procedures will cover all situations. No set of procedures will guarantee safety. No set of procedures can be a substitute for adaptive, intelligent human beings, and no set of procedures should ever be followed blindly. Procedures should be followed 'thoughtfully'. Procedures do not create safety, people do. This can be seen in situations where 'violation' of a procedure, accompanied by sound mental models and some good thinking, has resulted in work being done safely and successfully. This is further supporting the conjecture that the human contribution plays a major role in the creation of safe work.

Another expectation I have regarding procedures is that the procedures need to be suitable for purpose. I feel this can be achieved by ensuring absolute clarity about which bits are important and which bits it is okay to apply a bit of adaptive thinking to.

To this end, we can build procedures that contain two types of components, or sections. I believe calling these 'framework' sections and 'critical' sections of the procedures works as a naming convention, providing both clarity and purpose to the elements. Even if you do not agree with the nomenclature, I hope you get the idea.

'Framework' sections attempt to align with the concept of 'freedom within a framework'. This is where work teams and/or individual workers decide how the work will be done within a framework. The framework is set with clear guidelines and context but limited information on exactly how to achieve the task. The workers need to get together and think about how to do the work, come up with a plan, and then execute it. They have control of the nitty-gritty and this helps them think, adapt, make decisions, and act in alignment with the intent of the procedure.

'Critical' sections of procedures are critical in the sense that they are a step within a task that must be completed in a very specific way, accurately and repeatedly each and every time the task is done. The idea behind the critical sections is that failure to follow the critical steps that they contain in the way prescribed could result in a fatality or other similar significant incident.

Trying to get the balance right between framework and critical sections maximises the idea of promoting thinking – of providing that 'freedom within a framework' approach where and when it makes sense.

A much-needed leadership skill is to be consistent in our expectations regarding where the balance lies concerning compliance to procedures. We also have to be clear on how we feel about adaptation outside of the procedural requirements. This can be done through our behaviours and what we attend to.

Clear expectations should also be set regarding the accuracy of procedures. If procedures are not accurate, or if a procedure does not adequately cover some critical step or concept, then those expected to use the procedure may adapt and make up their own way of doing the task anyway. This clearly may not be a good thing in some circumstances. Therefore, to maximise the likelihood that procedures are accurate, we need to have a process that asks the following sorts of questions when reviewing procedures or any changes within them:

■ What is the value to safety of this procedure or its proposed change?
■ What is the value to the end user of this procedure or its proposed change?

- Will this procedure or its proposed change help us in the journey to 'always getting it right'?
- How does this procedure or its proposed change add or reduce complexity here?

KEY TAKEAWAY

When you understand your own expectations along with others' expectations, you have some access to their 'Why', to the context within which they operate and think. This will make life so much easier for everyone concerned than if you are not aware of those expectations.

Understands the Limitations and Use of Situational Awareness

It turns out that what we call situational awareness is very poor when compared to what is actually going on in any given situation.

A human cannot see the world in its entirety at any point in time. It is very much like looking through a pair of binoculars at a football match. We can see one little spot in quite a bit of detail but we cannot see what is going on at both ends of the field at the same time. Our view may move around the field but we cannot easily expand our field of view. In this way, we can only effectively perceive a small part of the world at any one time. We need to recognise that we are not capable of keeping an eye on everything in the workspace, nor, in fact, do we need to. The fact that we do not need to see everything at once is especially true as expertise develops in an individual. Think about driving. When we first get behind the steering wheel, we are overwhelmed by the number of things we need to keep an eye on and a side conversation about politics or religion is out of the question. Within a few years, however, we can drive and have an in-depth philosophical conversation at the same time with ease and skill. Two different people will also perceive the same work in the same work location at the same time in two different ways. This often comes up in post-workplace incident investigation interviews. We all see the same world differently.

It is not only in the visual field that we are not very good at perceiving the world. We are bombarded with signals in the workplace all of the time – there is always a lot going on. Not all of the signals, whether weak or strong, visible, tactile, or audible, are important and not all changes are of interest to us. We do, however, need to have systems and signals that tell us when things have changed so that we can process the meaning of the change and decide whether it is of interest to us or not. There are many ways this can happen and a lot of it is engineered like the haptics in an Apple Watch. Some of it is not, like the engine noise telling us when to change gear in our cars. Some of these triggers are very important, and we need to be very careful not to engineer out feedback at the expense of triggers for alerting us to a situation – situational awareness – for this reason. An example is in the modern motor car. We do not get the same sense of speed today that we used to get in cars in the 1970s. (I absolutely knew I was going fast at 120km/h in my 1968 Mini. I do not get that sense now in my Range Rover in 2021.) There is also so much more going on with the vehicle dashboard now than then. In my Mini Minor, I had a speedometer, temperature, and oil pressure gauge and that was it. In the Range Rover, I am now overwhelmed with computer controls, buttons all over the steering wheel, a radio that tells me what song is playing, a moving map that tells me where to go, and a tonne more seemingly arbitrary and disconnected stuff that I usually don't use or understand. Although I must confess that I don't think I ever looked at the oil pressure gauge in my Mini.

Pulling strongly from *The Invisible Gorilla* by Chabris and Simons can help people understand the limits of their situational awareness. If you have not seen the invisible gorilla clip, check it out on YouTube. We generally see far less of the world around us than we think we do. We suffer from not only visual blindness to our surroundings but also inattentional deafness. We tend not to see the things we do not expect to see, and we do not hear the things we do not expect to hear. Although we do actually hear them and see them, we just do not process the data into information. This is why telling people to be situationally aware is pointless – it is just like telling people to be safe or to pay attention at work. You need to be very specific about what aspects of the workplace you want people to focus on as people can only focus on a few specific things at a time.

Even given these limitations, situational awareness is practised everyday by everybody – whether they consciously think about it or not. The difference you can create in your business is that you can help your people understand the limits they have as human beings in this space. And that we all have a few tricks up our sleeves to assist us in understanding our workplace surroundings. As an example, we can learn what to look out for, rather than try to look out for everything.

'What is going on here?' is a fantastic question to ask about situational awareness during a field leadership conversation or when we sit down as a team to do a TBRA (Task-Based-Risk-Assessment) prior to starting a job. It is an open question and it can illicit responses that can pick up concepts such as sensemaking, mindfulness, situational awareness, drift, chronic unease, outcome bias, inattention blindness, focus, curiosity, observation skills, and seeing both strong and weak signals, just to mention a few that come to mind. Many of these ideas relate to the concept of 'mental models'.

A mental model is basically a theoretical blueprint of what is happening, what will happen next, and what could happen in the future. A shared mental model is simply a mental model that is shared – a mental model that is pretty much the same across a group of people. A mental model is in other words a representation of the aspects of the situation of interest that the individual or team is aware of – Situational Awareness.

It is a good idea to spend a lot of time on mental models with your people. It is a subject that can be discussed during task planning, during field leadership conversations, and during after-task post-mortems. These post-mortems are when, after a job is done, you talk about whether or not the task progressed in alignment with the mental model the team had of it before the task started, and what we can learn from it.

Encouraging teams to discuss their mental models as they go about their tasks, especially when something intrudes or changes during the task, can also help a lot. Get the team to get into the habit of stopping and talking about what has changed and how it may impact the way they think the job is being done – recognising that change can grab us unawares and that we need to reassess our mental model when that happens. This idea is closely related to resilience, the ability to create foresight to anticipate the impact of changes.

Having an appropriate level of worry about the possibility of failure is a worthy message to train people on as you explore the issues associated with situational awareness. So is trying to be reluctant to draw conclusions and to be sensitive to experiences and expertise, particularly so for the frontline operators. In short, having chronic unease is a lens through which you can operate – being mindful of what *could* happen – in a positive or negative sense.

Always be on the lookout for weak signals that may point to problems in the near future. Encourage people to stop and resolve differences within the team regarding their mental models rather than just ploughing on regardless of things changing and being different from the original plan.

At pre-shift or at pre-task meetings you can talk about the work and the way it is seen (the team's mental model). Then explicitly talk about what needs to go right, what could go wrong, and what plans the team may need to have in place to manage those things.

Get people to recognise that each time a routine is undertaken, it is done in a slightly different way and it pays to stop and think about what might be different, what might happen that could be different than before. Then plan for that eventuality.

To help create a mindful approach to mental models you can use field leadership conversations to stop and simply look around in a workplace with the worker and discuss what you all 'see' – this encourages the individual or the team to look and see beyond the mundane. The aim is to develop the skill of seeing differently, not just seeing more. This activity also aids skill development in sensemaking. Sensemaking is especially useful when we face an upset in our work, an encroachment into the normal way of doing a task. It helps in these situations to see things differently and hence resolve the problem more effectively and easily.

When you are in the field, you can easily create a conversation aimed at exploring situational awareness based on this broad set of questions:

- What assumptions are we making about how the work is, as compared to how we thought it was, when we wrote the procedure?
- Does everyone know their responsibilities and activities for this job?
- What do they think will go right? Wrong?
- What do they think about the procedures?
- Do they have plans for what to do if it all turns south?
- Are they seeing any drift? Any changes in the way they are doing the task compared to how they have done it in the past?
- What happens normally that may influence the ability to follow the procedures?
- What do they have to keep an eye on in the environment of the task? What level of situational awareness is needed, and on what, specifically?

To help people build the appropriate level of situational awareness along with their individual and group mental models, it is worth touching on the topic of foresight with your people. Foresight is a very difficult thing for people to become good at. Spend time working with and talking with your people, helping them understand that the workplace is not 100% ordered and that order breaks down sometimes. This state needs to be accepted as the way things are and that sometimes the things we do and have always done do not work out the same way each time. This means that we need to maintain a sense of chronic unease – even for those things for which we feel we have ultimate control and familiarity. Recognising this also drives us to be ready for an upset and be ready to adapt to the given situation as needed – to bounce back in the face of adversity.

Foresight is all about the 'what if?' It's the chronic unease and resilience engineering piece.

Resilience engineering as it relates to the individual and situational awareness is all about being able to respond to events, even before they manifest, as well as monitoring on-going developments and anticipating future opportunities and threats. Importantly, it's the skill of learning from successes and failures. Resilient performance is all about what people actually do. Viewing it this way, it can be taught. Observing weak signals, having the right sense of chronic unease, a lack of trust in the effectiveness of risk controls, and bouncing back from the face of adversity before anything hits the fan is an important ingredient in the recipe for creating safe work.

We need to look for resilient performance on a daily basis. We need to attempt to understand it, learn from it, celebrate it, and share it. A measure of resilient performance is the recognition in the work team that they have a good understanding of the now – their mental model of the work – and that they also have some foresight on what could happen next. They know what needs to go right, what could go wrong, what to look out for to indicate that things are possibly going wrong. They have a plan.

All of these techniques help people and teams to generate the right level of situational awareness in the workplace as they control the risks.

An Example

I was in a large mobile equipment workshop with the supervisor of the area. We had just entered the workshop and I asked the supervisor to stop and look around. I asked him what he saw and what was of interest to him. He responded by saying the obvious things like the fact that the individual truck bays were demarcated with chains and tags, that the workshop was clean and tidy, that all the mobile equipment tag out lock out boards were in place and being used. I asked if there was anything else of interest that he could see. I asked him what he thought about a maintainer working on a Caterpillar 793 truck directly in front of us. He noted that the maintainer was standing on top of the tire. He was about 3.6 metres above the ground. I pushed further about the harness, lanyard, and fall height. Until I pointed them out, he hadn't seen any issues. But then he saw the fact that the lanyard the maintainer was using was a fall arrest lanyard and was attached to an eyebolt on the top of the wheel guard at the level of the maintainer's feet. The maintainer would have firmly hit the ground before the harness was of any use. This was very obvious to me as an issue when I walked into the workshop but it was not seen by the maintenance supervisor until it was pointed out to him. It was only then that he clearly saw the issue. This is a classic example of situational awareness at work. We all have filters and we all see things differently.

KEY TAKEAWAY

We cannot see everything. It is not one of the skills a human can have, so do not expect it from yourself or your team. Instead, teach them how to see, how to think about what could go right and what could go wrong and to have the right level of chronic unease.

Doing

Listens Generously

My belief is that listening is the most important skill a leader can possess. The ability to listen and to be seen as a generous listener can set a leader apart from others. I have often heard people using a leader's listening ability as a surrogate for their ability as a leader.

Generous listening is an art. Listening is a skill. Listening is not always easy. However, listening is one of those things that can change the way you are seen by others and can directly impact your leadership effectiveness. How you listen also plays a big part in how we interpret what we hear. This can make a big impact on the effectiveness of task assignment or when we are sharing mental models and talking about our expectations and risk controls. How we think about our listening can significantly impact and improve our ability to listen. This idea is called meta-listening – that which lies beyond listening. The 'how' of listening if you will.

This meta-listening – listening to how we listen – improves our ability to listen. Meta-listening is a practice that we can adopt as a reflective approach to help us be more effective and impactful listeners. We also need to help other people understand how they listen and how they can keep improving their listening.

Our listening is an access to understanding. The more effectively we listen, the more we get to understand the world around us. Our listening creates our reality – the better we listen, the more we get a sense of the mental models of those around us, which feeds our reality. Listening is a tool by which we filter sounds and noises and how we select what to absorb into our knowledge base and what to ignore as background noise. Because it is a filter, it does not always filter in the way we anticipate it to. This leads, as we discussed in the previous section on situational awareness, to inattentional deafness, which can mean we hear but do not respond to what we hear. We heard but we did not listen.

Generous listening is much more than simply being quiet whilst others talk – it is all about focus and curiosity. Generous listening may also be described as active listening, or unselfish listening, or perhaps even disinterested listening. It is all about focussing your being on what is being said. Generous listening is not getting ready to respond. It is not thinking about something else like what's for dinner. Nor is it thinking 'I have heard all this before'. It is about listening not only with your ears, but with your eyes, your body, and your mind. 'Listening to understand' is another way of describing it. It is a super-important skill to have as a leader, and as an individual generally.

If you 'get' why generous listening is such an important tool, then you should explore how to get better at it. A good way to improve our listening as a leader is to always allow others to speak first. This has the result of encouraging others to express their views and opinions first and not simply respond to what you may have to say as their leader. Letting others speak first may result, if you are listening generously, in improving the quality of what it is that you subsequently say. Leaders speaking last shows that they care, that they are interested, and that they find what others have to say is important.

I came across a great way to encourage generous listening through a great consultancy called JMJ Associates here in Perth. They were helping me and my team develop our coaching skills. The idea they shared with us was about having a listening coach. A listening coach's job is to listen to our listening. It works best during meetings that involve conversations and interactions rather than in a presentation style meeting. You need to first find a volunteer to be the listening coach. The role of the listening coach is to listen to the people and to give feedback on how effectively the participants listened during the meeting. A listening coach looks for a number of things:

- The degree to which people are building on others' words, or go off on a tangent.
- The amount of time people give others who want to say something.
- The body language of the listener, including eye contact, body position, and focus.
- The number and intent of interruptions.
- Generally, whether the team members are listening generously or stingily.

In summary, listening is a skill that can be learnt and practised. My call to action is for each of you, the reader, to carry out a little post-mortem on your listening at least three times a day for the next week. During each post-mortem, think about the conversations you have just had and ask yourself how you listened and explore whether you could do anything differently next time to have listened more generously. Generous listening is all about listening with the desire to really understand what is being said.

An Example

Just having a listening coach changed the way our leadership team operated during its monthly meetings. We had one of the General Managers play a role, in addition to simply being a member of the team, of being the listening coach. I noticed that after the team all had a go at the role and got used to the fact that their listening was being listened to by one of their peers, team members seemed to pause before talking, attempted to build on each other's topics. This was a big departure from what we used to do, which often was to give the floor to the person who interrupted most or who was the loudest. Over a few months, the flavour of our meetings changed and I believe we all learnt how to listen better as a result.

KEY TAKEAWAY

Thinking about our listening improves our ability to listen.

Plans Work Using Risk Intelligence

As I mentioned in the introduction, I am using the term Risk Intelligence in a broad sense rather than the tight definition of experts like Dylan Evans. *Essentials of Safety* attempts to encapsulate some of the concepts and ideas of the

Efficiency-Thoroughness Trade-Off (ETTO), Risk Intelligence itself, and having a suitable wariness for the effectiveness of controls and chronic unease. Task planning is such an important part of the creation of safe work. It is important not only with respect to the high-level long-term and short-term planning that a planning function may do but also for the day-to-day, and especially the task-by-task, planning that individuals attend to all day every day.

Fundamental to people being capable of doing the low-level task-by-task planning is an understanding that we cannot do everything perfectly, all the time. If we want everything to be perfect every time, then we are probably going to be very inefficient doing the task. I could spend three days cleaning my car, for example, but that is probably not the best use of my time. There is always a trade-off between efficiency and thoroughness in the tasks we undertake, and task planning needs to take this into consideration. Hollnagel calls this trade-off the ETTO principle. You cannot achieve excellence in both efficiency and thoroughness at the same time. It is not possible. There is always a balance. We can help people understand the ETTO principle by getting them to talk about when and how they apply the ETTO principle in their daily lives outside of work. We can then talk about how it is being applied as they attack their daily work-related activities. Once you think about it a bit, we all ETTO all the time.

Imagine doing your Christmas shopping. You could create an active spreadsheet, including the people you plan to give presents to, the options for presents for each recipient, and then a series of columns with the various providers of said gifts sorted automatically by quality, availability, and price. Once this is done, you could apply an algorithm that balances these variables with an overall maximum budget spend for the year. This is well on the 'thorough' side of the ETTO. On the 'efficiency' side, however, you could create a simple list of names and present ideas, nip into your local shopping centre, wander around for a few hours, and grab whatever you see that fits the bill. It's more efficient, but you could also blow your budget or end up with sub-quality items. It is always a balance. This is the ETTO principle at work (Figure 1.3).

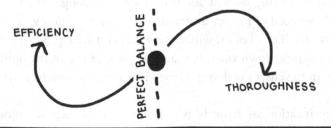

Figure 1.3   My version of the ETTO balance.

It is the balance of 'time to think' and 'time to do', or in the words, of Marquet 'blue work' and 'red work', that is at the heart of ETTOing. Helping people get that right is something that you need to push strongly. This is also a reminder that one of our tasks as leaders is not only to help people understand how to do their jobs but to arm them with the skills they need to do things that surround their jobs, like planning, handling uncertainty, managing adaptive problems, etc.

You can also talk about ETTO when we are in the field as leaders having field leadership conversations. We should have in our minds, as we have these conversations, that ETTOing can be a common driver of procedural shortcuts and procedural drift – heading towards efficiency compared to thoroughness in the application of the procedure to do the work.

You could talk with your people about the fact that due to ETTO, and just generally because we are all human beings, we will not get it right all the time. People need to expect failure and then to see what they can do to learn from it. Seeing failure as a learning experience rather than as a deflating experience is certainly the better attitude to have.

We all need to expect failure in ourselves. We also need to expect failure in the tools and equipment we are using. We all need to expect failure in everyone around us, including our leaders. They are, after all, human beings just like the rest of us.

Having an appropriate level of chronic unease also fits in with task planning and this expectation of failure nicely as well. That entails always having a level of alertness about what could go wrong. Having the mentality that everything going along swimmingly cannot last. Encourage people to try to keep an eye on things and have a back-up plan if things go south – i.e. anticipating a failure and already knowing what they are going to do to handle it.

If there is one guarantee in the world of safety and work, it is that people will fail. If someone goes up onto a scaffold with tools and equipment, they will drop some of them at some point in time. This is why we put up drop zones around scaffolding that is being worked on. Things *will* go wrong. When we drive our cars around, we need to be very aware of what others are doing. We know that other drivers, just like all of us, will sometimes get it wrong. We need to always look in the mirrors, down the side roads we pass, at the traffic lights, at what pedestrians and cyclists are doing, always planning to ourselves; what we will do if ...

The identification of hazards is a pretty critical step to understanding the situation in which we find ourselves as we perform a task at work. The identification of these hazards is not something, however, that we do once at the start of a task. It is a process that continues all the way through an activity. The work of planning to control risks never goes away. In order to plan for how to do a task, people need to develop risk awareness, and then maintain it throughout the task.

Developing risk awareness requires work. Over time and with practice, operators can achieve a high-level awareness of risk – a sense of what is going on around them and if anything dangerous is developing. I am reminded here of James Reason's idea of *error awareness* and *chronic unease* as well as Dylan Evans's concept of *Risk Intelligence.*

We can help our people develop the much-needed risk awareness we require. Allow them to spend enough time understanding their equipment and systems – looking at vulnerabilities, understanding risk profiles and critical controls as they plan the task – and then give them sufficient time and skills to monitor any changes and/or drift during the task itself, so they are always being mindful during the work. The usual suspects of situational awareness, mindfulness, skilful observations, and mental models come up again for us during task planning.

We need to touch on Risk Intelligence here also as it is all about our ability to estimate risk probabilities accurately. We all know that we are not good at estimating risk and we are also not very good as individuals in understanding the limits of our own knowledge. This is what Risk Intelligence is all about. We tend to jump to answers in order to look good or to say what the easy answer is. We should rather think about what we know about the issue, whether those bits of information make the event more or less likely, and by how much it impacts the risk. Then we should check whether this hunch makes sense and how strongly it is being impacted by availability bias. Only then should we make an estimate of the probability of something going right, or the probability of something going wrong. There are a number of methods to teach Risk Intelligence, including asking Fermi questions – questions that drive an improvement in estimating skills. This is useful when thinking about moving unknowns-knowns into known-knowns.

I am reminded of the Donald Rumsfeld stuff here:

> There are known knowns; there are things we know that we know.
> There are known unknowns; that is to say, there are things that we
> now know we don't know. But there are also unknown unknowns;
> there are things we do not know we don't know.

A classic example of a Fermi question that I have used is asking a group how many piano tuners there are in the local state or provincial capital. Normally, there is nobody in the group who has the exact answer but through some lateral thinking they can all get a pretty good estimate, especially when averaged. To do this, they can generally recall an estimate of the population. Then think about what percentage may own tuneable pianos. They then assume that one piano tuner, working five days a week and taking half a day to tune one piano, can easily estimate how many piano tuners there are. A quick review of the internet can get you the answer of course but that is cheating. Doing this helps the individuals get better at problem solving and improves their Risk Intelligence. The trick is to help them realise that by using the knowledge that they already possess, they can solve problems that at first glance seem intractable. We have found that this skill is very handy when attempting to understand the risks associated with complex tasks during task assignments.

Risk intelligent people have an ability to access bits of information somewhere in the backblocks of their minds that may be relevant, even if at first glance it does not seem obviously related to the problem at hand. It is true that many people cannot recognise the gaps in their own knowledge. They do not know what they do not know and they do not know that they do not know what they do not know – often they think they do know when they, in fact, do not know. I think I got that right.

Just like in all fields of endeavour, when we are trying to understand the level of situational awareness required for a task, it is unreasonable for a system to attempt to demand efficiency *and* thoroughness in the same breath. To help, we can build the systems that take this into consideration and hence do not put too many demands on the human to be aware of everything, all of the time. The way this can be done is to apply an ETTO filter to system development and changes. Try to make the system thorough enough such that the business is confident that the necessary and sufficient conditions will be created so that the system will deliver its objective. At the same time, we want to aim for the level of efficiency such that the system does not drive unnecessary resource use or requires people to have an unachievable level of situational awareness.

Most businesses are not High Reliability Organisations (HROs). Some do their utmost to progress in that direction and unfortunately only a few are crystal clear that a High Reliability Organisation is a journey and not a destination.

The standout area of focus for many on this journey is related to a preoccupation with failure and managing the unexpected.

When things are going well, leaders should worry. When things are not going so well, leaders should worry. Leaders need not be obsessed with what could go wrong; they just need to be preoccupied with it. This preoccupation with failure is sometimes called chronic unease.

We need to remember that having a good run with safety in the past bears absolutely no relationship to what may happen today or tomorrow. The absence of a failure, incident, or injury means *only* that there has not been a failure, incident, or injury – it does not denote anything else.

I first read about chronic unease in James Reason's books and loved the idea – as long as it is not taken to extremes, which some leaders have done in the past. Chronic unease is all about a healthy scepticism about whether stuff is going to be okay or not. I have heard the phrase 'wariness of risk controls' and 'vigilance' popping up in references about chronic unease and I recall a discussion in one of Reason's books about 'feral vigilance' used by the then Western Mining Corporation (now BHP), which also points to a constant lookout for what could go wrong implied by chronic unease. So, what does the behaviour or action of a leader look like if they possess chronic unease and how does that affect safety?

Leaders can show chronic unease by asking questions in order to encourage their teams to question the way they work. Leaders can encourage their teams to question the accuracy of the procedures they are expected to use. They can ask questions in order to encourage their teams to challenge any normalisation of deviance and drift. They can ask questions in order to understand for themselves what is driving any differences between Work-As-Done, Work-As-Normal, and Work-As-Written.

In keeping with the concept of wariness of risk control or chronic unease, one of the ways we can ensure we maintain a preoccupation with failure is that we can take on a systems perspective that tells us we must look beyond the individual mistake or 'error' and understand the underlying structures, culture, leadership, and system interrelationships that create the required conditions for a failure, incident, or injury to emerge. We should encourage people to have sufficient unease such that they approach each day as if something will go wrong and then plan for it.

When in the field, we should always ask questions such as what is going on here that is different?; why do you think it is different this time?; what can we do about it, to manage it?; and have we normalised the difference as we have seen it before without any adverse outcomes?

As a leadership team as well as individuals in the workplace, we need to track small failures and weak signals. Differentiating between noise and signal is most assuredly not an easy thing to do. It is our role as leaders to supply the ongoing effort needed to see the signals and help drive the application of adaptive efforts in order for things to go right as well as to stave off catastrophe. We need to keep a very close eye on attempts to simplify processes to make sure we are not over-simplifying things, and to keep a close eye on the moving locations of expertise. Simplicity can hide unwanted, unanticipated, unacceptable actions and activities, which could increase the likelihood of an adverse outcome. It is much easier to identify large signals such as changes. Be wary of these also. We should view change – in just about anything from the legislative environment, legislator's focus, technology and economic swings and roundabouts, production pressures, through to changes in equipment, tools, people, or level of complexity – as something that needs to be extremely well understood and managed.

Simply put, we should strive to ensure our systems result in HRO-style behaviours by our leaders. One example of this is with one of the system elements that can drive lead team routines mentioned earlier in the chapter. When you review or modify procedures for accuracy and then approve them, the system can drive you to always ask a few simple questions:

- Does the procedure (still) include mechanisms to make sure the job goes right?
- Has it simplified beyond clarity or sense?
- Has the review had effective involvement of the end user along with suitable expertise?

Another is in audit processes. When you are internally or externally audited, request the auditors ask a selection of leaders and frontline workers about how they believe the systems being audited aid progress towards an HRO-style operation. Welcome audits of your systems as they are methods that you can use to check yourselves and make sure you have not got too much complexity, too much complicatedness, or too much simplicity.

An Example

A client recently told me a story about a process they used in an attempt to simplify their workplace procedures and the planning that went into it. This project was not only aimed at increasing the level of simplicity but also about a serious reduction in the number of procedures and systems they had in their business. She told me that they were initially worried that the reduction in number and complexity of procedures could result in a lack of direction and an increased risk of incidents and injuries. She was also worried that they would get it wrong with respect to eliminating important bits and leaving requirements in the procedures and systems that were simply clutter. When I asked how she managed all of these diverse needs, she explained that they took a long time to get the planning right and were driven by trying to get the ETTO balance right as they approached the problem. They also tried to take the advice of Jim Wetherbee out of his book *Controlling Risk*. He talks about asking two questions: as leaders in the business, they asked themselves whether the system change they were planning left the system, along with all its procedures accurate or not. They also asked whether those that needed to use the procedures also felt that the changed procedures were accurate. Wetherbee goes on to explain what accurate means: 'Accurate includes being effective and representative of the organisation's collective wisdom (including the end-users) on the best way to accomplish the task'. The client went on to explain that applying this filter to decisions and changes resulted in getting the balance right between efficiency and thoroughness and ending up with what they felt was a balanced system. Was it perfect? Absolutely not. But it was a great deal better than before.

KEY TAKEAWAY

We, our systems, our people, our procedures, our plant and equipment, our culture, and the way we all work are not perfect. Keep a weather eye out for trouble and steer away from it before it hits you, especially as you do any planning activities. It is always about getting the balance right: efficiency versus thoroughness; chronic unease versus lackadaisical carelessness; absolute confidence versus wariness of the effectiveness of controls. Getting the balance right requires leadership, conversation, and having time to think (blue work).

Controls Risk

Controlling risk is a vast field – so vast that one could write a book on this topic alone. This is, in fact, what Jim Wetherbee along with many other fine authors have done. All I can hope to achieve here is to highlight the elements of controlling risk that do not always get the amount of attention I feel they deserve. This is not to in any way belittle any of the other methods of controlling the risk of course.

I also want to highlight the differences here in terms of managing risk and controlling risk. It is the management's task to manage risk, and it is the workers' tasks to control risk. I will leave the creation of the systems and the high level of risk management out of this section. It is covered in other parts of the book.

In this context, controlling risk is a balance of people trying to keep in mind several different things; why they are doing what they're doing; what their level of situational awareness is; what their level of hazard and risk awareness is; what their mental models are; what they put into their planning activities; the risk control measures they choose (hierarchy of control for example); their expectations regarding success and failure; preserving options; being mindful; and of course what tools and equipment, procedures, and systems they need to use.

The elements of controlling risks I mainly want to highlight are:

- Trigger steps.
- Critical steps.
- Preserving options.
- Telegraphing deliberate action.
- Shared space/safety signs.
- Risk Intelligence.
- Critical control verifications.
- Procedural compliance/adaptation.
- Drift.

The first two topics that I want to explore, which I feel are really important in controlling risk, are both more about what we need to do to make sure everything goes right rather than minimising things going wrong. These important topics are 'trigger steps' and 'critical steps'. It's not so much whether they are clearly described in a procedure or task-based-risk-assessment, but in that the team doing the work very clearly know them and are watching out for them as they do their work.

The difference between a trigger step and a critical step is subtle but important. A trigger step is one that has immediate consequences. Whether you get it right or whether you get it wrong, there is no going back after the trigger step. Once you have cracked the egg into the soup, you cannot get it back into its shell. It is like crossing the Rubicon – the point of no return. A critical step, meanwhile, is one that we must get right each and every time. If we do not get a critical step right each and every time, it could result in someone getting seriously hurt or killed. Both trigger and critical steps need to be identified prior to starting the work. They should be talked about as a part of the sharing of mental models and monitored throughout a task.

When decisions are made during the task, especially related to changes in the work method – such as on-going task planning – we can make sure the teams have preserved their options for action in the event of things going pear-shaped. An example is always making sure that there is a second exit option if something goes wrong, or another control that can be implemented easily if one fails. In the outside-of-work world it could look like keeping an eye on the sides of the road so you know where you could go and what you could do if a car coming the other way swerves into your lane. Another example is riding out wide on your bicycle when you pass parked cars in case someone decides to opens a car door on you.

It is useful to spend time thinking about, and talking with leaders about, how to set up workplaces that are 'error tolerant' and whether it is possible to 'fail safely'. These ideas/ideals are about the workplace and are super-important goals. They also recognise that humans will fail. And we need to recognise that we cannot build systems that are human-proof. It is also worth mentioning here that systems are not things that are 'applied' to people, but that systems include people. This tends to be forgotten when we talk about system development. We should go a bit deeper than simply 'humans will fail' and start thinking along the lines of:

> In addition to setting up the workplace for success, what about the
> human bit – the person doing the work, or the work crew around the
> person doing the work. What can they do to help maximise getting
> it right, minimise mistakes, slips, lapses etc?

Jim Wetherbee and David Marquet gave me the answer. I believe that their discussions about taking 'deliberate action' and 'telegraphing' actions are brilliant and something that should be explored and shared.

The ideas of taking deliberate action and telegraphing actions are so similar that I felt it was okay to combine them and talk about telegraphing deliberate action.

Telegraphing deliberate action is all about getting into the habit of not only stopping and thinking about what you are about to do, but also physically pausing just before you do the action and, at the same time, verbalising your intent to yourself and to those around you.

So what does telegraphing deliberate action look like in anger? Here is a story that may explain it in a bit more detail:

Scenario: An Elevated Work Platform (EWP) operator is manoeuvring an EWP so that the basket, with the operator and a workmate in it, moves away from a hot furnace that they are working on. Moving the EWP in the wrong direction may result in the people in the basket of the EWP receiving serious burns. Adopting 'telegraphing deliberate action', the EWP operator holds his hand over the 'B' lever and announces to no one in particular 'I am moving lever "A" in order to lower the basket'. The other worker, who is standing next to the EWP operator in the basket, sees the hand movement and hears the intention of the operator. He immediately alerts the EWP operator of the discrepancy. Alternatively, the EWP operator himself, upon saying lever A and seeing his hand over lever B, stops the action and corrects it himself. Either way, the potential burn is avoided and safe work is achieved. This is a fairly typical mining or construction industry example, but I hope you can see the potential for using 'telegraphing deliberate action' in your workplace – whether that be a hospital ward, operating theatre, maintenance workshop, office, or mine site.

When I talked with operators and leaders about this idea, they tended to say that it made sense and they could see how it could help, but they generally felt that it would not work so well for them personally. Some said it would feel weird to go on chattering as they did their work. They said they would feel uncomfortable doing it. This led to what I feel is the biggest problem with getting to an effective use of telegraphing deliberate action – we do not generally think out aloud as we do work. We do not share our ideas as we get things done in a normal work situation. In fact, the 'safe to speak up' concept that is often driven by businesses and often seen in industries may not extend to the cultural 'need to be seen to know what you are doing and just getting on with it' way of working. The benefits of 'telegraphing deliberate action' only manifest when people are actually speaking up as they are doing their work. This may potentially present some difficulties during implementation. One way around this is for leaders to practice telegraphing deliberate action in their normal daily activities, in their meetings, in their workplace interactions. They can then get into the habit of talking through what they are thinking and doing and encourage their teams to listen and help poke holes in their logic. Thinking out aloud is a classic example of this. Of course, imperative in establishing 'telegraphing deliberate action' is helping others understand the 'Why' of the activity. If they get it, if they deeply understand why telegraphing deliberate action may help them minimise mistakes and help them undertake safe work, they will do it. If they don't get it, they won't do it.

What I love about the idea is that even if other operators are not present, the application of 'telegraphing deliberate action' benefits the operator themselves because the act of pausing and the vocalisation of intent forces the individual to be mindful of the present situation and what is trying to be achieved. It gives us another opportunity to get it right. And getting it right in the first place is much more satisfying than doing an investigation afterwards.

You need to make sure the message is *not* 'we telegraph deliberate action for the benefit of the observer, the leader doing some field leadership conversation, or the boss'. The consistently shared message needs to be that 'telegraphing deliberate action is purely for the benefit of the person doing the work'.

Jim Wetherbee says this about telegraphing action:

'When the practice of telegraphing actions becomes automatic between crew members, the operating effectiveness of the team improves dramatically. When executed properly, this practice contributes to error-free operations, allowing the team to achieve better performance, with higher-quality results'.

This is a reminder that telegraphing deliberate action is not only about safety but is also about production and quality.

Whether you are a manager, an operator, a nurse, a doctor, or an engineer telegraphing deliberate action makes a huge difference to the level of mindfulness and situational awareness in the workplace and helps you to get it right the first time. So, it is not only about reducing mistakes but it is also very much about operational excellence. It also helps keep our workmates up to speed with what it is we are about to do – or at least up to date with what we *think* we are about to do. By sharing our intention, we are sharing our mental model of what the work situation is and what we are about to do within it.

I have a request from you. Do a micro-experiment (to borrow a Dekker idea). Go away and have a play with the practice of telegraphing deliberate action. Talk to people about it. Get them to have a play with it. Do it yourself for a while and see how it feels – there will be benefits.

A great example of telegraphing deliberate action, although he does not call it as such is in Richard Mullender's *Dispelling the Myths and Rediscovering the Lost Art of Listening*. Mullender talks about how police drivers are trained to describe what they see and do as they drive to get better situational awareness and become better drivers. Part of an example he uses sounds like this: 'I am driving at 40mph. In the distance I can see a road joining me from the left and the junction is clear. There is no-one behind me and there are two cars approaching'. And a bit later:

> As I turn into the junction I can see there is a car coming towards me and there is a junction on my left hand side that is clear. I check my mirror and there is one car behind me travelling fast.

Making telegraphing deliberate action a routine prompted me to look closely at the impacts of routines themselves.

Even though having routines is a great way to be consistent, absent-minded slips are more of a problem for experienced people than they are for the newbie in the activity. Slips are more common during habitual activities undertaken by people who know what they are doing, and there is no doubt that 'routines' fall under the same banner of 'habitual activities'. Watch out for small departures from the routines and habits that you may have developed. Keep an eye on them and also on any small changes in what is going on.

As we establish routines, or changes to routines, we need to keep a very close eye on what is going on and what could go wrong, including anything we have not seen before that may pop up unexpectedly. Once again, always have an escape plan. Always have an option for when something unexpected happens. Never paint yourself into a corner. I remember when I used to ride my Ducati down the highway. I never sat the bike in the blind spot of a car or a truck. I always watched cars on side roads and had an opt-out strategy just in case some inattentive car driver did something unexpected.

How we set up a workplace can also give a false sense of safety and security. Safety signs are a classic example of this and one that I feel require a few words on their own. Safety signs can be important in some circumstances. However, more often they can be the bane of a leader's existence. They are routinely everywhere and usually add very little, if any, long-term benefit. There can also be an overconfidence related to the effectiveness and efficacy of them as risk controls. I suggest instigating two activities to address these issues: a safety sign war, and a risk control wariness campaign based on the idea of shared space.

Firstly, let's talk about a safety sign war.

Establish regular workplace inspections to make sure you get rid of any sneaky posters and safety signs – ones that add zero value such as: (on mirrors) 'You are looking at the person responsible for safety', 'Be careful – Do not get hurt today', 'Safety is YOUR responsibility', 'Remember to Take Care, Be safe', 'Electricity can kill', 'Work safely', a sign on a railway crossing that reads

'Vehicles must not become immobilised on this crossing', and this really smart one I saw on a hot water tap: 'CAUTION – The water in this tap is hot and can burn you'. Imagine … hot water coming out of a hot water tap – who'd have thought!

I have experienced, as we removed signs in a workplace, people telling us that they did not know that half of them were actually there. They had become such a part of the furniture and were no longer even being seen. People also said that they were glad to be treated as adult humans instead of as children.

Try getting rid of the safety messages on front gate electric smart signs that talk about the number of days since the last injury and the injury frequency numbers from it as well. Everybody will be happy with that one. These are absolutely not useful measures of whether work is being done safely or not. They are just easy to measure.

To further expand beyond your war on safety signs, take a look at other controls from the perspective of shared space.

'Shared space' is a way of thinking that was derived from a traffic-management philosophy created by Hans Monderman, who hails from the Netherlands. Even though he was focussed on traffic management, the idea is equally useful in a workplace. It originally involved the removal of traffic controls such as stop signs, traffic lights, speed limits, pedestrian crossings, barriers and gates, caution and warning signs, and painted lines from a section of the road. The idea was to increase the unease of road users and encourage mutual respect and communication – either direct or indirect, including eye contact and visual cues – in order for all road and intersection users to interact safely and effectively.

I am not encouraging the wholesale removal of controls. I am merely stating that you need to look before an incident, during our day-to-day or normal work, as well as after an incident to check if you have overdone the risk controls. You do not want to dumb down your people by encouraging them to be reliant on signage and controls without clearly understanding them and their intent.

Encouraging people to think before they act is a good thing. As a leader, ask yourself if allowing people to have some level of discomfort and encouragement to think before they act is such a bad thing. It is not.

As usual, a balance needs to be struck. I often hear complaints that the practice of 'safety' means wrapping people up in cotton wool, with a long list of procedures and processes that cause workers to stop thinking for themselves about what could go wrong and what they must do to prevent it from going wrong. This is essentially where the thinking behind shared space intersects with the designing and the creating of safe work. We want to provide enough information to help the workers make decisions but not overload them with 'protective' controls. Once again, it is all about getting the balance right. I also talked about this with respect to framework and critical sections of procedures earlier. We also need a prompt that helps us think about what level of overprotection played a part in an incident or an LNW review.

We all know that work is not risk-free. We also all know that it is the task of leaders in our business to apply risk management and it is the task of frontline workers to apply risk controls.

The typical view in people's mind is that risk is all about hazards and danger, but for those who understand the concept of risk, danger is only one side of the risk coin, the other being opportunity and success.

I believe that one of the secrets to risk management from the perspective of a leader lies in the concept of Risk Intelligence and in setting up risk assessment processes – helping people to understand what risk is actually all about.

Treating risk assessment processes for fatal/material/catastrophic risks in a very similar way to that associated with risk assessments prior to the creation of procedures and the like is sensible and beneficial in terms of creating safe work. The two processes are different in terms of outcomes but the inputs are pretty much the same.

The other difference is that, in bowties and material/fatal risk registers, you highlight the critical contributing causes and also the critical controls. These are controls that must be in place each and every time the task is done, or else the hazard may manifest. You should always try to keep the critical controls to a couple and then make them very specific. Then you need to carry out verification activities on these critical controls in the field as a part of field leadership conversations. One common term for this is a CCV (Critical Control Verification).

The purpose, or why, of the CCV is to verify that the critical controls from fatal and material risks are in place and effective. CCV's are a very conversation-based approach and provide a strong level of assurance for the lead team in the business with respect to critical control implementation.

Before we start identifying causes, consequences, probabilities, and controls in a risk assessment workshop, we need to take the team through the risk issue with the intent of getting the team to build a shared mental model of the system and its behaviour as well as the mental model of the specific element of risk issue itself. This step is routinely missed in risk workshops I have been subjected to.

It is important to do a lot of work before and during any risk assessment workshop. This includes training the participants in Risk Intelligence and reminding the participants that all risk is subjective and is interpreted through the lenses we all have of our own experiences, expertise, imagination, Risk Intelligence, biases, heuristics, recent incidents, perspectives, etc. We then need to fill the risk assessment workshop with people who provide diversity in their thinking, experience, and knowledge associated with the risk issue and of course some 'real' people. I use this term in incident investigations/learning studies as well. These are the people who actually do the work – the operator, maintainer, nurse, plumber, construction worker, etc. Real people know the Work-As-Normal – how the work really gets done.

As for Risk Intelligence, it is the ability to estimate probabilities accurately, whether we are talking about the probabilities of various events occurring in our lives, such as a car accident, a workplace event, or the probability that some piece of information we've just come across is actually true. We often have to make educated guesses about such things, but 50 years of research in the psychology of judgment and decision-making show that most people are not very good at doing so. Many people, for example, tend to overestimate their chances of winning the lottery, while they underestimate the probability that they will get cancer at some time in their life.

In addition to the use of Fermi questions (that we previously discussed), during risk assessment workshops, one way of improving Risk Intelligence is to gather wider information on the subject, or to look for things that may disprove our view, not for things that support our view. This helps us remove some of the confirmation bias.

So why is Risk Intelligence an essential skill for our leaders? Because we want to help them learn the ability to assess situations in which normal work is done and also explore what could go wrong (and right) and the likelihood of that happening. In order to be able to do this, they need a reasonable level of Risk Intelligence. It is this view of risk that we want people to think differently about.

A simple approach to improving Risk Intelligence is to expose ourselves to a greater diversity of opinion, and especially to seek out views that are opposed to our own. This manifests in the workplace as the requirement that task-based risk assessments are best completed as a team, involving those new to the team and others as much as possible. The young and the newbies often ask some great questions and make a real difference to the risk assessment process and outcomes.

It is a good idea to remind the risk assessment participants to keep an eye on their availability heuristic as this can impact how they think about and perceive the risk.

The availability heuristic works well enough when we have to estimate the probability of things that are entirely within the realm of our personal experience, such as missing the bus or finding a ten-dollar note on the footpath. But when we're gauging the likelihood of things that are reported in the media, the correlation between ease of recall and likelihood breaks down, and the availability heuristic leads to biased estimates. TV news and social media present us with rare and dramatic disasters that we might never otherwise see, such as plane crashes, tsunamis, terrorist attacks, and so on. The images sear themselves into our memories and are recalled rather too easily. If we rely on the availability heuristic when estimating the probability of such dangers, we will tend to think they are more common than they really are.

What can we do to avoid being led astray by the availability heuristic and imagination inflation? The most obvious remedy is simply to be cautious when estimating the probability of dramatic events. If images of such events come easily to mind, we can ask ourselves if it is because we have personally experienced many of them or because we have read about them on the internet or seen them on TV, or been exposed to them through some social media platform. Likewise, we can ask ourselves if such events loom large because they have occurred in our lives or because we have previously allowed our imagination to give them full rein.

The availability heuristic leads us both to overestimate the probability of dramatic events and to underestimate the likelihood of situations that are not so easy to picture.

Another one we need to watch during risk discussions, especially in risk assessments, is probability neglect. This results when a hazard stirs a strong emotion in the individual. What happens then is those involved in the risk discussion will tend to overestimate the likelihood of the hazard manifesting. They will assess risk in a way that reflects their wish to avoid the hazard manifesting.

A conversation about controlling risk would not be complete without a quick discussion on the hierarchy of control and engineering risks out of the equation. I make the assumption that you are aware of the concept of the hierarchy of control. My point here is to not forget it. A final area that is often misunderstood is the use of procedures. I have talked about them under the expectations section of this chapter in detail but it is worth briefly revisiting that for a bit now. Procedures are rarely written in such a way as to be foolproof. They require their users to adapt to the unique situation in which they find themselves at work. Work practices also drift over time, with the way the work is being done today being different from how it used to be done. People pick up ways of working that may be easier or quicker and over time that becomes the way the work gets done. Drift can also appear as procedural drift in Work-As-Written, where we have tweaked procedures over time due to periodic reviews or changes after workplace incidents to the detriment of the level of control of the risks. We can end up with a procedure that misses the mark on critical controls and may actually add risk to a task.

To help with understanding and keeping in mind what needs to go right and what might go wrong, we can prompt people to ask themselves questions similar to those below:

- How do I normally get this trigger step or critical step right?
- Do I have a plan to make sure everything continues to go right?
- Am I doing enough telegraphing deliberate action?
- What is in the line-of-fire that can seriously bite me?
- How do I work within the housekeeping as it currently is?
- What is going on around me – including any simultaneous operations (SIMOPS: SIMultaneous OPerations) and will I impact it, or will it impact me?
- What is going on in relation to Material or Fatal Risks – critical controls that I should be considering? Is anything I am doing related to a Bow-Tie or Material Risk/Catastrophic Risk?
- Are there any anomalies in the workplace that do not appear to have been there before? What stands out as different?
- Do I have prompts to help me understand and be aware of the critical risks?
- What devices or elements provide me with alarms and warnings when danger is imminent?
- Do I have a plan if something does go wrong?
- What if … ?
- What usually goes right here, but may not today?
- What will happen to me and the team if we do not control the risks?

KEY TAKEAWAY

Controlling risk is all about maintaining a balance. You need to consider: why people are doing what they're doing; what their level of situational awareness is; what the level of risk awareness is; what their mental models are; what they put into their planning activities; the risk control measures they choose (hierarchy of control for example); their expectations regarding failures; preserving options; being mindful; and, of course, what tools and equipment, procedures, and systems they need to use. And this is, of course, all led through authentic leadership.

Applies a Non-Directive Coaching Style to Interactions

As you are developing your approach to leadership and the development of your people through these essential elements, I hope you realise that it must be supported by an approach that will help your people be the best they can be. One way of doing this is to provide coaching. The mistake leaders often make is assuming leaders know how to coach. This is not always true. Leaders are often good mentors, good teachers, good answerers, and instruction givers, but not all are naturally good at coaching. We need to help leaders understand what coaching is and how to do it effectively.

When businesses are interested in helping leaders become good coaches, they often start by bringing an external consultant in once a year or so and have them coach the leaders directly. This approach can absolutely make a difference, quite often one that lasts a reasonable time. But it is not a sustainable solution. If, on the other hand, you were to bring in a coach and had them develop, mentor, and coach your leaders to be great coaches for their teams, you can end up with a sustainable and on-tap resource within your business that can make a huge difference over the longer term.

It is a bit like the old saying 'Give a man a fish and you feed him for a day; teach a man to fish and you feed him for a lifetime'. I have seen so many leaders become much more effective in all areas of their work once they understand the impact they can have as a coach in addition to their impact as a leader and a manager.

I am sure many of you have seen several forms of coaching models over the years and thought that most of them are fine. Some are more complicated than others, but if you apply any coaching model that puts the player first (and makes that your intent during a coaching session), you will generally not go too far wrong. Having said that, as I mentioned above, I have a favourite. It is simple, powerful, beneficial, and really fun to use. It is what I currently use with many clients.

I chose to align my coaching style to that described in Myles Downey's work. It entails the understanding that coaching is neither teaching nor mentoring, and understanding that coaching is instead all about learning, growing, and raising awareness. You can choose your own coaching method of course, but I believe that Downey's explanations and examples provide excellent clarity and advice and are a great start to any excellence-in-coaching journey.

As I am sure you are aware there are many elements to coaching and the one I want to share with you – as it is so powerful in practice – is a model called the GROW model and it works like this:

GOAL (The G in GROW) is about establishing what the desired outcome of the coaching conversation is all about. It is very much driven by the player with the coach prompting the player with a series of questions to help the player think about exactly what it is that they seek assistance with for the specific coaching session. It quite often is not the broad-brush initial statement that ends up being the goal of the session. The secret to a successful coaching session is in the questions the coach asks. These help to focus the player on a very specific topic for the goal of the coaching session.

REALITY (The R in GROW) is concerned with achieving the most accurate picture of the current state of play that it is possible to achieve. The reality part of the coaching session is always directly related to the goal of the session and it is here that the coach really helps the player see what the world looks and feels like for them. It is in this phase that the coach's skills of listening to understand, following interest, generating understanding, and providing feedback and choice really come to the fore.

During this stage, there should not be any analysis, no offering of bright ideas, no suggestions of 'how I would do it is …', and no jumping to conclusions.

OPTIONS (The O in GROW) is about what can be done – what is possible for the player to do. The intention here is to draw out a list of all that is possible without judgement or evaluation. Questions such as 'Which of those would you like to pursue first?' 'So what could you do differently?' 'What else?' 'Anything else?' 'What else can you think of?' are the food for the coach at this stage. The Options stage is usually directly related to an aspect of the Reality stage when the player has identified an element that is not quite how they would like it. For this reason, it is worth pointing out here that the GROW stages are not always linear. Sometimes, it helps to move backwards and forwards a bit between the phases as the conversation grows. With practice, it becomes more of a style of conversation than a rigorously applied process.

WRAP-UP (Or Will-do, the W in GROW) aims to select the most appropriate option from the Option stage of the coaching session and agree the next steps. In this phase, the coach's intention is to gain commitment from the player for an action. Things like 'So tell me, what, exactly, are you going to do?' appear here. This is where one of the options becomes a reality that the player can take away and do. Coaching is for the purpose of improving performance, and so it must result in tangible ways of being or doing that change the view, approach, or actions of the player in order to be effective. This is an essential component and the player needs to get it and to commit to going away and doing it.

To give it some visual perspective, this is how I usually draw the GROW model. The idea being that it is not a forced flow but a flicking back and forth between Goal, Reality, Options, and Wrap-up (Figure 1.4).

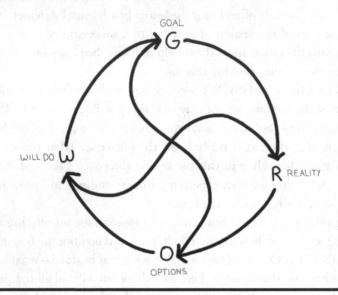

Figure 1.4   The GROW model.

Coaching (and using a coaching style) needs to be recognised as an essential skill for all leaders. It needs to be taught, supported, nurtured, and practised over a long period of time.

Once leaders have an ability to apply true non-directive coaching when they want to, they can then, and often do, apply a range of behaviours that could easily be described as a coaching style in their management activities on a day-to-day basis. They ask – not tell. They apply intent-based leadership. They limit 'instructing', and they limit 'command and control' behaviour. This is what is usually meant when we talk about a leader applying a coaching style to their leadership or management activities and approaches.

Coaches and leaders using a coaching style know that the answers to problems lie within the capacity and expertise of the player, not with the coach or with the leader. This ties the concept of using a coaching style in management and leadership to intent-based leadership that David Marquet so masterfully discusses. By this, I mean that the answers to problems and issues usually lie within the expertise and knowledge of the worker, not within the expertise and knowledge of the leader.

Doing the thinking for players may seem like the easy solution to help people be the best they can be. This is usually not true. And is often a waste of time – in the longer term. People learn much more effectively if we get them to think by asking great questions and listening to the answers. You need to work at the level of thinking – thinking about how we think about our thinking – as thinking goes to talking which goes to actions. That is where the rubber hits the road with the player. We need to aim to improve players' performance by helping them improve their thinking through our coaching. After all, people are paid to think so we need to help them get better at it.

Using a coach to help us combine our leadership and coaching styles with a thorough understanding of your 'Why' can really help us drive towards a state of passionate and authentic leadership. This is not rocket science, nor is it complex or complicated. It is simply about focussing on a few things that all work well on their own but are hugely powerful when combined. It is that fluidity that can be strived for, easily moving between managing, coaching, mentoring, and a spot of telling.

A bit of an aside about feedback. Feedback, whether a part of mentoring, coaching, or simply a part of managing, is only useful if it has been asked for and if it is authentic. I value the concept of feedback and am very careful to give it and ask for it. I make sure I do not pump people up based on the outcome of things they do, but rather for the effort they put into the process – remaining quite often, somewhat silent on the outcome. I am also very mindful of the impact that feedback can have on an individual's state of mind. If feedback is given purely for the purpose of helping others, in a similar way that coaching is intended, you will not go too far wrong. It can be very useful in helping people realise that they can have a growth mindset, and can then grow and have potential. And we can help them do that through coaching. If they already feel that they are the best thing since sliced bread, this can tend to support a fixed mindset. On that note, please give me some feedback about this book (ian.long @raeda.com.au). I welcome it.

The principal difference between coaching and mentoring is that with mentoring many of the answers lie with the mentor. With coaching, nearly all of the answers lie with the player. If you have trouble with this during a coaching session, try to remember the maxim: 'let expertise lie where expertise lies'.

There is a bit of guidance that I have found useful during difficult bits of coaching conversations, or for that matter, difficult conversations during mentoring or general leading and managing activities. It is the 'Crucial Conversations' tool described by Kerry Patterson et al. This employs seven simple steps and is highly effective in helping the process move along and be effective:

1. Start with the heart (i.e. empathy and positive intent):
   - This ties in with being an authentic leader as well.
2. Stay in dialogue:
   - No arguments here. Always build on the other's words.
3. Make it safe:
   - Included in here is authentic and generous listening.
4. Don't get hooked by emotion (or hook them):
   - Keep focussed on the end game, refuse to get into an emotive argument – back off, listen, pause, and *then* respond.
5. Agree a mutual purpose:
   - Find the common ground about what you are trying to achieve here.
6. Separate facts from the story:
   - Recognise and call it when you are no longer in the world of facts and have moved into the world of opinion and stories.
7. Agree a clear action plan:
   - Make a mutual decision and then work out what that looks like.

In the end, we can practice, promote, and pursue excellence in coaching, using a coaching style, and mentoring our people for the purpose of helping everybody be the best they can be.

Here are some examples of some questions that I have found useful when preparing for a coaching conversation:

Goal

What is it you would like to discuss?

What would you like to achieve today?

What would you like from this session?

What would need to happen for you to walk away feeling that this time was well spent?

If I could grant you a wish for this session, what would it be?

What would you like to be different when you leave this session?

What would you like to happen that is not happening now, or what would
  you like not to happen that is happening now?

What will be of real value to you?

What do you want?

Reality

In relation to … (the Goal), what is happening at the moment?

What has been working? What has not?

What have you tried so far?

Tell me about …

Who or what are you doing this for?

What's happening?

What is the most difficult part of this task?

What did you like most about the way you accomplished this task?

How do you know that this is accurate?

When does this happen?

How often does this happen? Be as precise as possible.

What effect does this have?

How have you verified, or would you verify, that this is so?

What other factors are relevant?

What is their perception of the situation?

What does that mean?

In what way?

What do you understand by x? What don't you understand?

Did you observe anything in particular to make you think he/she was happy
  with your feedback?

What are the anticipated consequences of x?

What do you notice as you look at x, or consider y?

At your best, what qualities, attributes, capabilities do you bring to the
  situation?

What is your understanding of the current situation?

Options

If you could turn back the clock and try that again, what would it look like
  or sound like?

What could you do to change the situation?

What alternatives are there to that approach?

Tell me what possibilities for action you see. Do not worry about whether they are realistic at this stage.

What approach or actions have you seen used, or used yourself, in similar circumstances?

Who might be able to help?

Would you like suggestions from me?

Which option do you like the most?

What are the benefits and pitfalls of these options?

Which options are of interest to you?

Rate from 1–10 your level of interest in the practicality of each of these options.

What alternative possibilities can you consider?

What would success in this look like?

What do you feel most strongly about in this situation?

What stands out?

Tell me, in an ideal world, how would things be different?

What would it look like in x months or years?

What changes would you like to make?

If you could do it any way that you wanted, how would you go about accomplishing this task?

When have you succeeded in a challenge similar to this one?

How comfortable (confident) do you feel about doing x?

Where could you find the help you need to accomplish this task?

What would it take to make you feel more comfortable?

What do you really want?

Wrap-Up

What are the next steps?

Precisely when will you take them?

What might get in the way?

Do you need to log the steps in your diary?

What support do you need?

How and when will you enlist that support?

An Example

Karen Ross, an exceptional senior safety professional in BHP, who I have worked with for many years, kindly provided this example of the impact coaching and using a coaching style can have. She wrote: 'Such a simple concept, so how did the GROW model have such an impact on my life?

Well it all began about 10 years ago, when my leader (aka the author of this book, or Longy as he is generally known) started sprouting all this information about how this GROW model could change the way we interact with our people and this would lead to better safety outcomes. The concept was very easy to understand, but I was sceptical. At the time, I was studying risk management and believed risk management was the Holy Grail to improving safety.

Each month, Longy would come for a site visit and I was very quick to identify where I thought our issues were, but I was not following through on how to resolve them. Then I started to notice when I raised the issues with Longy, the conversations we had were very different. We talked about what it could look like, where we were now, the options we had to get there, and what I needed to do. Mind you, this was not without a few throw away lines made by me to Longy such as: "Stop doing that GROW shit on me again Longy".

I then thought that maybe there is something in this coaching. I started practising with my team, peers, general manager, and even the family did not escape from my GROW experiment. I was constantly surprised by how engaging and easy the conversations were. The beauty was that I did not need to know the answers, just how to ask the questions. I finally GOT IT, people noticed and it was a turning point in my career. I had learnt to ask questions differently.

Over the past 10 years I've had the privilege of sharing the GROW model with lots of people, who have all been surprised at how easy and effective it is. I'll be forever grateful for Longy's passion and patience to make sure I got it.'

KEY TAKEAWAY

You coach and adopt a coaching style in your leadership and management toolkit to help your people be the best they can be. It entails asking questions in a way that promotes thinking and learning. It is easy to learn, amazingly effective, and fun.

Has a Resilient Performance Approach to Systems Development

Before I talk about how we can use the ideas of resilience to help create and monitor the systems that help drive safe work, I will recap what resilience engineering is all about.

Resilience engineering, at least in terms of how it fits in with our current conversation, has four potentials of interest to us, and these need to be encouraged, measured, and talked about. These are the potentials to Respond, Monitor, Learn, and Anticipate:

Respond: Knowing what to do when trouble goes down or is about to go down.

Monitor: Knowing what to look for or being able to monitor things that could go wrong.

Learn: Knowing what has happened and being able to learn from the experience.

Anticipate: Knowing what to expect or being able to anticipate developments into the future.

We should think about resilience as we have our field leadership conversations and interactions. This means checking the resiliency of our procedures and teams doing the work. We should assess whether the teams have thought about what needs to go right, what could go wrong, are keeping an eye on what is going on as issues develop, and have plans to bounce back from the face of adversity into safe work without things going south. We need to identify resilient performance and celebrate it, understand it, and learn from it. In other words, we need to establish how much of Work-As-Normal represents resilient performance on a day-to-day basis.

Resilience is pretty much the same as saying we are coping with complexity and is all about successfully coping with the unexpected – including, and maybe especially, at the time before the unexpected materialises as an unintended outcome or incident.

Historically, we have tended to use the ideas of resilience to do three things: prevent something from going wrong; prevent things from getting worse; and recover from stuff that has gone wrong. I suggest trying to focus more on using resilience to help make sure stuff goes right in the first place rather than just on preventing things going wrong. This aligns with my view of what safety is. As I mentioned in the introduction, 'Safety' is about maximising things going right rather than an absence of things going wrong. It is about people in a system, not a system driving people. It is about maintaining the balance between thinking and doing.

To this end, I have thought about a slightly different focus for the explanations of what the four potentials are. The intent was not to change them but to focus them more positively:

Respond: Knowing what to do when things start moving away from going right.

Monitor: Knowing what to look for or being able to monitor things that need to be in place to ensure things go right.

Learn: Knowing what has happened to make things go right and being able to learn from the experience.

Anticipate: Knowing what to expect or being able to anticipate developments into the future.

The only real way of determining if resilience is present in a system or a workplace is to get out there and witness resilient performance on a day-to-day basis. Remember that resilient performance is always about balancing a trade-off between goals. It is to be found every day, if we only look for it and encourage it.

Specifically, we should look at our systems – by which I mean safety systems, procedures, leadership behaviours, individual and team thinking, and how they all interact – and explore whether we have set up a system that drives Responding, Monitoring, Learning, and Anticipating as work is being planned and undertaken. This is mainly done through field leadership conversations and whenever we review, audit, verify, or in any other way check the effectiveness, usefulness, and accuracy of our systems. Here I do not just mean the high-level corporate systems but also the low-level procedure and work instruction level of systems that are used each and every day in the workplace.

How we create and modify our safety systems over time always needs to have a lens of resilience engineering over it. You could say that we need to attain a culture of resilience, if there is such a thing.

System Safety as an idea or concept is not new. Neither is Process Safety, nor is Safety in Design. I will not spend time here digging into what each of these looks like. I will say, however, that what we need to do within these frameworks is to build in resilience early, as well as approaches that emphasise the handling of complexity and the elimination or control of hazards, allowing humans to fail safely. Do not wait until the failure has occurred to work out what tweaks we need to make to the system, but build the system to include that eventuality.

It is arrogant for us to believe that we can set up a system that will be completely resilient. Resilience lies in resilient performance. So what we should do is to try to set up a system that encourages people to be aware of weak signals and give them the skills to adapt to them, to make sure it continues to go right. This is how we can build resilience into our systems.

We want to include stuff that helps us understand when things are going well, or becoming brittle – before they break.

It is also worth mentioning here that systems are not things that are 'applied' to people, but that systems include people. This tends to be forgotten when we talk about system development.

In this section, I want to describe a possible system approach and how it got to look like it does.

When developing systems of work, or tweaking existing elements of our systems, we should consider how a number of different things and processes interact and/or act together when exposed to a number of different influences at the same time. In other words, we need to apply both systems thinking and complexity thinking, often simultaneously. It is often the interrelationships between the various elements of a system that, only when considered together, make any sense.

It is useful to summarise the drivers and intents of the different parts of a safety system and how they interact. This requires a considerable amount of work but is essential because the reasons behind those procedures are easily forgotten as people move on and times change.

Historical/corporate/collective memory such as 'I seem to recall we had an incident a number of years ago related to this stuff …' is not sufficient to retain learning. As we see in the paper by Fanta et al., the reliance on humans passing on learning to others is a faulty mechanism for retaining fear of catastrophe. Individuals will tend to remember details and incidents in their recent past, but individuals move on, and only a second-hand memory may continue. Instead of relying on this, we can build the information into the story behind the elements of the system. We can try to make these stories emotive, powerful, and visual to maximise learning for the current system users.

The systems need to be designed (well, they need to be intended to be designed) to allow people to fail safely. We know that it is not possible for 100% of our brilliant people to be 100% focussed and perform 100% accurately 100% of the time. There is one guarantee in life, and especially in the workplace, and that is that people will fail. Give someone a spanner and ask them to climb onto a scaffold, and they will drop it at some point in time. Give someone a syringe and there will be stick injuries every now and again. Look at babies learning to walk – a clear example to show that failure is a critical part of learning.

As we create the systems that we intend to use, we need to fully recognise that Work-As-Done does not always equal Work-As-Written and that procedures by their very nature cannot handle the unanticipated things that pop up in the real world. We can also intentionally and actively integrate into the systems, both HF/E (Human Factors/Ergonomics) and a detailed study of Work-As-Normal – how the real world works.

A useful lens through which to view the world as we are building systems is to apply some thinking to what could or should go on. It is useful to explore the links and relationships between functions within a system and how variability in one function can impact others. The Safety Oscillation Model described below was built after considering such a review.

In a nutshell, I broke down a system into a set of functions that relate to each other and then I attempted to draw how they interacted. I then explored sources of variability within each function and explored how it may play a part in other functions within the system. In many ways, the resultant model talks to how we think the systems would work as a completely interrelated system rather than simply a whole lot of separate bits. This played a key role in developing the Safety Oscillations Model detail.

Safety Oscillations Model

A model that greatly helped me understand and tweak a model 'safety' system was built after a review of *Resilience Engineering: Concepts and Precepts*, Edited by Erik Hollnagel, David D. Woods, and Nancy Leveson. CRC Press 2006. Specifically, it was a figure describing a simplified model of the dynamic behind the space shuttle Columbia loss. It is also used in a paper by Leveson in greater detail. I have generalised it to talk about the creation of safe work. I adapted it and it is used both to understand how the system elements influence other elements and how it can be used by safety professionals and business leaders to keep an eye on things and help them emphasise or de-emphasise activities for the purposes of getting it right. I wrote a simple paper on it and it is reproduced here:

The Oscillations of Safety in modern, complex workplaces

We tend to measure safety outcomes as a surrogate for safety. In the model, I have called a difference between 'safe outcome' as an

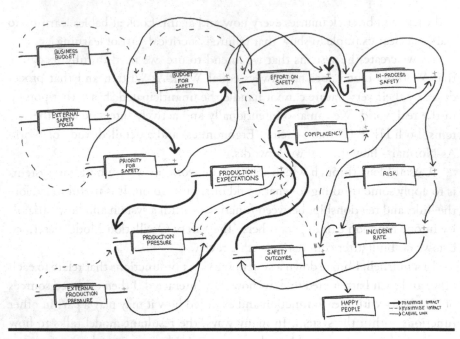

Figure 1.5   The Safety Oscillation Model.

outcome and 'in-process safety' – stuff that makes things go right on a day-to-day basis (Figure 1.5).

There are a number of inputs to safety outcomes and in-process safety. Some are closely coupled and some more distant. Some of the more closely coupled include: effort on safety, production pressure, budget for safety, and priorities for safety. Some of the more distant couplings include: business budget, external safety focus, external production drivers and risk view. I have attempted to capture them in the flow diagram above, which highlights the buttons that can be pushed to drive things in the direction we want and away from things we do not want.

Note: The darker lines represent something that we could focus on – maximise/strengthen, and the dashed lines represent things that we could minimise – push back against. The simple solid lines represent simply how one thing will/can lead to another.

So, let's start with the 'Safe Outcomes'. If 'Safe Outcomes' goes up ('Safety' gets better such as a decrease in injuries), represented by a + sign on the right of the 'Safe Outcomes' box, then the laps start:

An increase in 'Safe Outcomes' can lead to three different things. One is 'Happy People' – something we should maximise. Another is an increase in 'Complacency' as everything is going well. This is clearly a link we want to de-emphasise as much as we can. The third link is from an increase in 'Safe Outcomes' to an increase in 'Production Expectations' as 'Safety' is humming along nicely. This is another thing we want to minimise and de-emphasise.

Firstly let's follow the increase in 'Complacency' link. An increase in 'Complacency' can lead to a reduction in 'Effort for Safety' – which leads to a decrease in 'In-Process Safety' and then an increase in 'Risk', an increase in 'Incident Rate' and hence a reduction in 'Safe Outcomes'. Continuing the right-hand loop for now, a decrease in 'Safe Outcomes' can lead to a decrease in 'Complacency' – which we want to encourage. This can result, with some help/pushing, in an increase in 'Effort for Safety' and hence an increase in 'In-Process Safety', a subsequent lowering of 'Risk', a reduction in 'Incident Rate' and hence an increase in 'Safe Outcomes'. Which is where it all started. This oscillation is never ending but also has other inputs and impacts.

Another previously mentioned result from an increase in 'Safe Outcomes' is an increase in 'Production Expectations'. This needs to be minimised as we talked about earlier. An increase in 'Production Expectations' leads to an increase in 'Production Pressure'. This leads to a decrease in 'Effort for Safety' and around the right-hand loop we go again.

If we now look at one of the more distant couplings. 'Business Budget' for example. An increase in 'Business Budget' will result in, with the right emphasis and push, in an increase in 'Budget for Safety'. This will result in an increase in the 'Effort for Safety' which results in an increase in 'In-Process Safety' and 'Safe Outcome'. Around and around the mulberry bush we go again. You can work out the rest.

We can use a tool from resilience engineering to keep an eye on things. If we build our safety capability to monitor and detect the weak signals within each 'element' so that we can alert when the element gets close to a failure boundary and move the emphasis away from the boundary. We can then direct effort to either emphasise or de-emphasise the link to the next element and hence impact

'In-Process Safety' and 'Safe Outcome' directly and/or indirectly by maximising and minimising the links as required'.

KEY TAKEAWAY

Spending time knowing what to do when things start moving away from going right, knowing what to look for or being able to monitor things that need to be in place to ensure things go right, knowing what has happened and being able to learn from the experience, and knowing what to expect or being able to anticipate developments into the future, and then including that knowledge into your systems, leadership routines and conversations will go a long way to ensure we get it right, each and every time.

Adopts an Authentic Leadership Approach When Leading Others

I, along with many of you I am sure, worked out long ago that the best and most effective leadership style by a long way is authentic leadership. Authentic leadership is best described as leaders leading from who they really are. A firm and unalterable goal should be that your leaders are truly authentic. Authentic leaders know who they are and why they do what they do. They act according to their values and principles. They care for their people. They have good relationships with peers and those they lead. They are real and come across as genuine. They have a growth mindset – admitting their mistakes. They recognise that leadership is an improvisational art, know that leadership is not about power, persuasion, and personality, and they truly believe it is all about helping others be the best they can be. They are on the balcony and on the dance floor, as and when needed. They adapt. They all know that leadership starts with them. And that adopting the leadership styles of others just does not cut the mustard. At the same time, they know that they can learn from others, and then work out what that means for themselves. We need to always bear in mind, of course, that these are aspirational goals, and in reality, we have leaders who sit all over the spectrum of these ideals.

A concept/idea well worth exploring in relation to authentic leadership is that of intent-based leadership. Here we are talking about giving control to those who have the information – giving the decision-making authority to those with the expertise. Intent-based leadership is about people setting the context and intent of the work and allowing those doing the work to tell you how they intend to do the task. It is an extremely powerful way of building trust. The idea is that if people have the competencies to do the job and have the clarity of what is needed, then we should let them be in control of how the work is to be done. This fits in with the framework and critical sections of procedures and freedom within a framework concept nicely. Of course, you can simply teach other people how to undertake intent-based leadership, but the Holy Grail is when you practice intent-based leadership yourself. 'Do as I do', not 'do as I say'. An essential mode of operation for any leader in using any leadership style is allowing, encouraging, and fostering time for their people to not only *do* but also *think* about leadership and what their roles and tasks are all about. Once again, it is all about getting the balance right. Marquet talks about the difference between 'doing' time and 'thinking' time as 'red work', or getting things done, and 'blue work', thinking about stuff. Thinking about stuff includes thinking about what the next piece of red work is all about. Great leadership is all about getting the balance right as I have said numerous times before. Maybe we should call the balance 'purple' work. Maybe not.

It is also well worth remembering that each individual is just that, an individual. And each leader is also an individual. You need to make sure that as you help your leaders develop, your expectations regarding leadership and leadership style suit the leader's individual characteristics. Do not try to create a set of leadership clones. That is a recipe for disaster. Having said that, there are some basics that I believe should be covered in any leadership development programme.

As I have just mentioned, to aid in people's leadership journey, a leadership development programme should be capable of being adapted to suit the needs and wants of the individual leader. A common underlying skill is that of maintaining our focus on our presence – how we come across. We should aim to be authentically calm, confident, communicative, consultative, controlled, curious, comfortable, clear, and caring.

In addition to this, any leadership development programme you have can, and in my opinion should, explore the following topics:

- Inquisitive mindset.
- Generous listening.
- Acceptance of being proven wrong as well as being proven right.
- Chronic unease.
- Decision-making.
- Intent-based leadership.
- Focus on achieving excellence compared to preventing failure.

- Deference to expertise.
- Seeing failure as learning.
- Asking compared to telling.
- Coaching, including as a leadership style.
- Providing soon, certain and positive feedback.
- Adaptive problem solving.
- The *Essentials of Safety.*

There is a lot of things in that list, and in many ways they are all pulling on our leaders all the time. It, again, needs to be a balance (Figure 1.6).

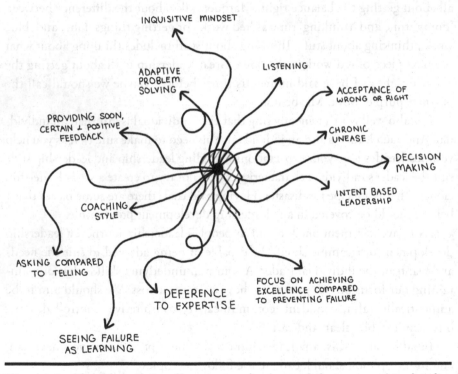

Figure 1.6 The things that pull on us as a leader, and that we need to develop in ourselves.

We need to always remember that leadership is something we do with our people, not to our people. We should encourage our leaders to discover for themselves what leadership characteristics they need to apply. There is no one-size-fits-all – no simple leadership formula that can be applied to all and sundry. We must not require our leaders to read exemplars of great leaders and then attempt to emulate them. We need to teach them to understand who they really are so that they can exhibit their own brand of authentic leadership. One thing we can do is to remind them strongly about being consistent and doing what they say they are going to do – i.e. having integrity. They can also be encouraged to talk about what they do well, and what they do not do so well.

I acknowledge that although it takes effort, extensive practice, and engagement, leadership can be taught. And we should try to help people learn how to explore their own leadership. Leadership is all about relationship. Relationship with the concept of leadership; the relationships that we develop as we lead; our relationship with mindset (growth or fixed); the relationship with our coach (you *need* at least one); and the relationship with our confidence, or overconfidence.

Although it is easier for our leaders to assume that those who report to them have the right level of leadership competence, we need to verify this through conversation and consultation.

In a nutshell, we should encourage, promote, and coach our leaders to be authentically themselves, only more so.

KEY TAKEAWAY

Leadership can be taught. Leadership is not about copying someone else's style. We want people to be themselves only more so, setting context, and by being authentic, helping their peers and teams be the best they can be.

Conclusion

Maintaining a balance is key to safety. Actually, you need to maintain a heap of different and interwoven balances. What I hope to have achieved in this chapter is the explanation and exploration of 12 of these 'things to balance', split into 2 fundamental groups of *Thinking* and *Doing* – with 6 in each. If we can get these six *thinking* elements and six *doing* elements in place, effective and balanced, we will be well on the way to creating great and safe work.

In summary then, the essentials of safety are a set of individual characteristics, distinctions, attributes, or traits that permeate through the workforce at all levels. As described in detail in this chapter, the above essentials or 'essential elements' talk to each viewpoint of the *Individual*, *Leaders* and leadership, the *Systems* we use, and the *Cultures* of the workplace. It is evidenced by a state where, driven through strong relationships, everybody:

Thinking
- Understands their 'Why'.
- Chooses and displays their attitude.
- Adopts a growth mindset – including a learning mindset.
- Has a high level of understanding and curiosity about how work is actually done.
- Understands their own and others' expectations.
- Understands the limitations and use of situational awareness.

Doing
- Listens generously.
- Plans work using Risk Intelligence.
- Controls risk.
- Applies a non-directive coaching style to interactions.
- Has a Resilient Performance approach to systems development.
- Adopts an authentic leadership approach when leading others.

Using the above as a framework for thinking and doing – or behaving, individuals, whether workers, leaders, technical, or support people, regardless of their hierarchical level in the organisation, will be internally driven to do work, set up ways of working, including procedures and systems, behaviours, practices, processes, and routines that align with these elements. It is through the consistent and interrelated application of these essential elements that the workplace culture will manifest, the systems will be developed, leaders' behaviours will emerge, and individuals will thrive. In many ways, as we have seen in the chapter, all of the elements and all of the ways the elements can be considered are all about people being the best they can be. This should not be surprising as individuals are people, leaders are people, those who create systems and people and all the people together create the workplace cultures in which we all work.

Further Reading

Set up by essential element, what follows is a list of authors that I recommend you read to get further understanding of the ideas that helped me get the past, current, and future thinking that led to the element in question.

Understands Their 'Why'

Simon Sinek (29, 46); Amy Cuddy (69); Rob Goff and Gareth Jones (117)

Chooses and Displays Their Attitude

Stephen C. Lundin: John Christensen (32): David D. Woods (5, 75): Eric Hollnagel (34, 74. 75, 78, 82, 98); Andrew Hopkins (19, 27, 49, 50, 51, 83).

Adopts a Growth Mindset – including a Learning Mindset

J.G. Mahler (64); Todd Conklin (6, 68); Steven Poole (76); Erik Hollnagel (33, 74. 75, 78, 82, 98); James Reason (1, 38, 53, 56, 103); Tom Nichols (97); Steven Shorrock and Claire Williams (39); Sidney Dekker (5, 22, 33, 45, 66, 81, 84, 100, 109); Carol Dweck (57); Chesley B Sullenberger III (91); Art Kleiner, Jeffrey Schwartz, and Josie Thompson (112).

Has a High Level of Understanding and Curiosity about How Work Is Actually Done

Erik Hollnagel (34, 74. 75, 78, 82, 98); Ian Long (89); Todd Conklin (6, 68); Atul Gauwande (96).

Understands Their Own and Others' Expectations

Sidney Dekker (5, 22, 33, 45, 66, 81, 84, 100, 109); Erik Hollnagel (34, 74. 75, 78, 82, 98); Andrew Hopkins (19, 27, 49, 50, 51, 83); Daniel Kahneman (113); Art Kleiner, Jeffrey Schwartz, and Josie Thompson (112); Scott Snook (35); James Reason (1, 38, 53, 56, 103); Diane Vaughan (95); Corinne Bieder and Mathilde Bourrier (114); Jim Wetherbee (15); Adam Higginbotham (56); Edgar Schein (63); Peter Senge (101); Jake Chapman (92); Atul Gawande (96); Sidney Dekker (5, 22, 33, 45, 66, 81, 84, 100, 109); Todd Conklin (6, 68); David Woods (5, 75) ; Vaclav Fanta, Miroslav Salek, and Petr Sklenicka (30); Margaret Stringfellow (2); Nancy Leveson (25, 58).

Listens Generously

Myles Downey (24); John Whitmore (9); Carol Wilson (67); L. David Marquet (115).

Understands the Limitations and Use of Situational Awareness

Sidney Dekker (5, 22, 33, 45, 66, 81, 84, 100, 109); Erik Hollnagel (34, 74. 75, 78, 82, 98); James Reason (1, 38, 53, 56, 103); Amy Cuddy (69); David Rock(71); Dylan Evans(77); Rob Long (51, 73, 94); Karl Weick (54, 65); Sally

Maitlis (87); Christopher Chabris and Daniel Simons (104); Daved Barry and Stefan Meisiek (85); Florence Allard-Poesi (105); Andrew Hopkins (19, 27, 49, 50, 51, 83); Mike Lauder (41); Guy Walker, Neville Stanton and Paul Salmon (116); Daniel Kahneman(113).

Plans Work Using Risk Intelligence

Andrew Hopkins (19, 27, 49, 50, 51, 83); Dylan Evans (77); Tom Nichols (97); Erik Hollnagel (34, 74. 75, 78, 82, 98); L. David Marquet (115); Karl Weick and Kathleen Sutcliffe (54); Peter Senge (101); Sidney Dekker (5, 22, 33, 45, 66, 81, 84, 100, 109), James Reason (1, 38, 53, 56, 103).

Controls Risk

Thomas Fisher (18); Andrew Hopkins (19, 27, 49, 50, 51, 83); Rob Long (51, 73, 94); Sidney Dekker (5, 22, 33, 45, 66, 81, 84, 100, 109); Dylan Evans (77); Nancy Leveson (25, 57); Roy Fitzgerald (94); Erik Hollnagel (34, 74. 75, 78, 82, 98); James Reason (1, 38, 53, 56, 103); Diane Vaughan (95); Barry Turner(52); Ian Long (89): Jim Wetherbee (15); Karl Weick (54); Kathleen Sutcliffe (54, 65); Corinne Bieder and Mathilde Bourrier (114); L. David Marquet (115).

Applies a Non-Directive Coaching Style to Interactions

John Whitmore (9); L. David Marquet (115); Myles Downey (25); Max Landsberg (55); Ronald Heifetz (106); Sharon Parks (47); Carol Wilson (67); David Rock (71); Kerry Patterson, Joseph Grenny, Ron McMillan and Al Swizter (16); Carol Dweck (57); Ian Long (89).

Has a Resilient Performance Approach to Systems Development

Erik Hollnagel (34, 74. 75, 78, 82, 98); David Woods (5, 75); James Reason (1, 38, 53, 56, 103).

Adopts an Authentic Leadership Approach When Leading Others

Bill George (20); Simon Sinek (29, 46, 90); L. David Marquet (115); Rob Goff and Gareth Jones (117); Sharon Parks (47); Amy Cuddy (69); Sidney Dekker; (5, 22, 33, 45, 66, 81, 84, 100, 109) Joe MacInnis (17); Art Kleiner, Jeffrey Schwartz and Josie Thomson (112).

Chapter 2

Leaders' Perspectives: Practices and Routines

There are significant benefits to thinking differently about leadership.

Turn the Ship Around, **L. David Marquet**

There are significant benefits in then moving from' thinking' to 'doing'.

Ian Long

Introduction

It is fine to talk about 12 essential elements of safety in theory, but what does it all look like in the life and behaviours of an individual, especially if that individual is a leader? This chapter aims to answer that question. At the end of the day, the term 'leader' covers pretty much everyone in the workforce. It takes the position of practices, routines, and activities that a leader can do that will utilise each of the 12 *Essentials of Safety* elements.

Understands Their 'Why'

I seek to understand my 'Why' in order to give myself a clear direction, anchor and signpost for everything I do. This is to maximise the likelihood that 'what'

DOI: 10.1201/9781003181620-1

I do and 'how' I do it aligns with the reasons 'why' I do it. I refer to my 'Why' when I am asked whether I am interested in doing some work with a client. I talk about my 'Why' when people ask me what I do. I am very driven by my 'Why'. My 'Why' is

> To share ideas, concepts and practicalities in safety and leadership with as many people as will listen, so that people start to think differently and positively about the why, what and how of the things they do in both safety and leadership.

It is the reason I have spent a huge amount of time working on this book. It is what keeps me aligned and going forward. I truly believe that the outcome of the work I did working on this book will help people think differently about leadership and safety.

Teaching ourselves and our teams how to understand their 'Why' needs to be approached in a defined way. It is a mistake to simply ask people why they do what they do or to ask them what their 'Why' is. Many people who have not come across the concept of their 'Why' before will not really understand the question, or will understand it at a superficial level. In order to explore how a 'Why' can be built, I do not go any further than simply following the advice contained in *Start with Why* by Simon Sinek and *Find Your Why* by Simon Sinek, David Mead, and Peter Docker. The latter especially offers approaches for helping people find their 'Why'. It also offers great advice and process for helping teams find their 'team why'. They advocate people sharing stories with a partner. These can be about specific events that contain standout memories and maybe things that have occurred that are the significant moments of the past. Think about stories that have made the individual feel good, proud – stories that have made the biggest differences in their lives. During the sharing and conversations about these stories, a few themes or common elements will arise. These are often examples where there have been significant contributions that have resulted in an impact. After a few shared stories, conversations, and some soak time, it becomes pretty clear that these themes or common elements can be converted into a 'Why'. A 'Why' statement is in the form: 'To .... so that ....', where the first blank is a contribution and the second gap is an expression of the impact of the individual's contribution. This is the format at the start of each element in this chapter - to try to put the element in perspective of its 'Why'. There is more to understanding your 'Why' than this simple explanation, but this is enough for you to get your 'Why' with a bit of work and thinking. It is preferable to try to find your 'Why' with the help of a partner but it can be done on your own. That is how I developed mine. It took quite a long time and my 'Why' has gone through a few refinements since then. It is now in a form that inspires me to action. My 'Why' motivates me to give presentations that are powerful and meaningful – I hope! My 'Why' inspires me to sit here with a coffee in my local coffee shop each day and share my ideas and thoughts with you by writing this book. By the way, if you are ever in Perth, Australia, Community Coffee Co in Subiaco is simply the best.

There are a number of ways that you can establish routines to keep your 'Why' alive and relevant. It is important to regularly invest time in thinking about your 'Why'. Make it a part of what you do each day. Not for the purpose of changing it or updating it but as a driver of motivation and inspiration. Make talking about your 'Why' a part of who you are. Encourage your team members and peers to talk about their 'Why' and include it as a part of the routine performance assessment activities with your team.

After disconnecting from a Microsoft Teams link following a 'virtual' investigation training workshop, I got a text message from the Executive General Manager of the company I ran the workshop for. It read: 'Thanks Longy, well done mate, cheers'. My response was 'Thanks, I love sharing ideas with people. So it is hardly what I would call work'. This, as you can see is very aligned to my 'Why'. Other opportunities arise all the time. I also have a routine where I formally look at my 'Why' every six months and ask myself if it still looks and sounds how I want it to, whether the work I have been doing in the last six months aligns with it, and whether I have shared its ideas with others recently. On this last element, I have found that if I share my 'Why', it helps others understand what drives me and our relationship strengthens as a result. In a way I am sharing something that is personal, is all about me, and that helps a lot in relationship building.

I often run coaching sessions with leaders. I also do a lot of coach-the-coach conversations with budding, developing, and already good coaches. When doing this 'work', before discussing with them what they want out of the coaching session – the 'Goal' in GROW coaching model language - I explore why they want to be coached in the first place. If they do not know why they want to be coached, I question whether we should continue. I know that I am wasting everybody's time trying to coach someone who does not want to be coached, or who does not understand why they should be coached. If, on the other hand, the individual knows why they want to be coached, or why they want to get better at coaching, the coaching will be far more effective.

I was asked recently to help a close friend understand who they are in terms of what drives their approach to work. She had just been promoted at work and wanted to make sure her language and expectations were founded on a solid base. I suggested starting the process by her getting to nail down her 'Why'. We started by talking about examples of times in her past where she had felt strong, happy, really inspired, proud, and felt that she had really contributed to something meaningful. After sharing these stories with me and jotting them down on the whiteboard, we looked at what the common elements or themes were. These were ultimately translated into the contributions she has made and the impacts. From this, her 'Why' was pretty easy to extract. There were probably four versions on the white board at one time but it was not long before one stood out as really representing her 'Why'. She continues to talk about her 'Why' with her new team, especially when asked to talk about what she does.

Chooses and Displays Their Attitude

I seek to choose and display my attitude by exploring what task needs to be done, why it needs to be done, how it needs to be done, and most importantly, what my attitude towards it is. I choose how I am going to react to it *before* I start the task. I do this as it allows me to choose to approach it in a positive way, regardless of the work. This is extremely powerful in terms of my well-being and happiness when working as well as my effectiveness in completing the task.

Teaching ourselves and other people about choosing and displaying their attitude is not always easy. It does not lend itself to a dedicated workshop or a one-off training session. For leaders to learn about choosing and displaying their attitude, it is best taught by you being an example in your words and in your behaviour. Start by sharing examples, coaching, mentoring, and through your general leadership behaviour. This is a classic case of leading by example. And it is also a classic case of learning by example. If you want to make the idea of choosing and displaying your attitude a solid part of your leadership style and the development of others, I suggest reading *Fish* by Lundin, Paul, and Christensen, as a great place to start. It is a simple book in the parable, story style of *The Goal* and others of that ilk. If you don't want to read another book just now and want to focus on a simple way of teaching others about the importance of choosing and displaying their attitude, it can be achieved by establishing personal routines in your work behaviours where others see you attacking tasks in a manner that is consistent with choosing a positive attitude. Make sure you also do this when the task is mundane and is generally considered to be boring. Have conversations about the attitude you have chosen for a task. Talk about how you feel about the work you are doing. As you undertake a task that others do not like, mention why you are approaching it with a smile. This, of course, must be done authentically and not just by saying the words and forcing a smile. If you try to bluff them with a smile, you may well come across as untrustworthy and fake – your words may not match your body language.

I used to hate doing the monthly budget reconciliations and checking the nitty-gritty of accounts. It was an absolute drag and something I did not look forward to each month. The accounts people would hassle me. I would also hassle my managers to get theirs done, and I kept putting off the inevitable until the last possible minute. One day the poor accountant who got lumbered with helping my department sat down with me over a cup of coffee and talked through why she loved doing the month-end reconciliations with the department head to whom she was assigned. She said that it was not a glamourous part of the job and was sometimes frustrating when working with people like me, but when she managed to get through to us, the rewards were worth it. This approach helped her look forward to the monthly interactions in the positive hope that she would get a breakthrough in our understanding. This attitude grew and she started looking forward to the meetings and approaching them with a smile with a view and attitude of possibility. I still struggle to get highly excited about reconciling bank accounts for my accountant these days but I do approach it with a view that it is actually quite easy. It only takes a short time and really helps me understand where things stand financially.

The biggest change was when I took this idea of looking at things through a positive lens on board across a broader scale. This was especially true when I started spending a bit of time assessing whether a task to be performed aligns with my 'Why' or not. Reading and summarising well over a hundred books and papers as a part of the research for this book was at times hard work. I knew, however, that it aligned with my wish to share ideas with people. This attitude helped me formulate ideas and thinking and made the task so much easier. I found that I relished the time seeing what others think about and how aligned my thoughts were to theirs.

Adopts a Growth Mindset – including a Learning Mindset

Why do I seek to adopt a growth mindset? I do this in order to be in a position of recognising that I have so much more to learn and improve. This enables me to seek every opportunity I can to learn so that I can share and explore meaningful ideas with people.

How do we teach people about mindset and how do we encourage them to adopt a growth mindset? Whilst this may not appear to be easy to do, there are many things that we can do to promote the idea and to help people move from a fixed mindset to a growth mindset. We need to remember that we cannot simply turn off a fixed mindset and replace it with a growth mindset in ourselves or in others. We cannot just run a training session and tell everybody to start having a growth mindset and being open to start learning new things. It takes time and effort accompanied by workplace factors that support the change. Leaders' conversations, the systems that support training infrastructures, and the way we frame work can all make a difference. We need to start with ourselves as leaders. When we move to helping others, we need to start small so as to not overload people. Running a series of small workshops is a great way to start the thinking processes. You will need to give sufficient time to allow the participants to progress at their own speed. You can start by explaining the differences between a fixed mindset and a growth mindset. Include a discussion on why having a growth mindset may be a lot more useful to the participants than a fixed mindset. A useful approach is to have the participants think about how they have learned skills they currently have and how they have got to the point of being good at their current jobs. Getting to a position of skill and competence in a job will most likely be achieved through practice and persistence and not through endless training sessions. The idea is to bring the participants on a journey of discovery that they have learned a lot in their roles in the past and that they continue to learn things every day. The participants need to come to the answer themselves – that they can have a growth mindset and there is still a lot to learn.

There are a number of simple routines to help keep learning and having a growth mindset front of mind. Whenever your organisation suffers a workplace incident, instead of asking about who was responsible for the incident, ask instead 'What can we learn from this?' You can encourage other people to talk about their failures by talking about your own, and then talking about what you have done to learn from those failures. I do this frequently when talking about my last book, *Simplicity in Safety Investigations*. In that book, I talked about Work-As-Intended as a way of describing the way we think work is done via procedures. After extensive conversations and feedback from clients, along with my own experience of using Work-As-Intended during the facilitation of workplace incident investigations, I came to the conclusion that this was not quite the right language to use. In *Essentials of Safety,* and in conversations I now have with new clients, I talk about Work-As-Written in place of Work-As-Intended. This much more clearly expresses the original intent of the term. What I learnt from this was that I needed to spend sufficient time engaging with end-users, clients, and peers *before* creating new terms, especially ones that are critical to a process I am attempting to explore and share.

When thinking about a learning mindset, it pays to ask your team how they like to learn. You can then tweak your learning options to suit them. This will not only pander to their learning styles but will also definitely make the process more effective. You might think that the most effective method of learning is exactly how you have always done it. This does not mean the participants think that way, nor that it is the most effective method. Mixing up the learning approach can also generate learning benefits to participants in ways that you have not yet thought about.

Using the word 'yet' when talking about what you or someone else cannot currently do adds an element of growth mindset. For example, rather than saying 'I do not know how to use Microsoft Teams', we should say 'I do not know how to use Microsoft Teams yet'. This opens up a possibility for learning.

Another practice to elevate the conversations surrounding growth mindset is to help provide sufficient time for people to maintain the balance between thinking and doing. Allowing people the time to work through why they do what they do, what there may be to learn about a situation, and more specifically, how they choose to behave in a certain circumstance. This also aids in greater learning from any situation.

We should always challenge team members to push their skills envelopes. Both Carol Dweck (*Mindset*) and Angela Duckworth (*Grit*) contribute greatly to how we can have conversations about mindset and how it impacts our lives authentically.

Front-end-loading a leadership selection and development programme enables a growth mindset to be at the forefront of conversations. A client shared with me what they were doing when hiring new leaders into the organisation. It was based on the typical targeted selection approach where the candidate is asked to share examples from their previous work or life around a specific topic. An example from personal experience is: 'Tell me about a time when you saw an employee taking a procedural safety short cut and what you did about it?' The interviewer is looking for an explanation of the situation: what the background or context was, what the task was, what were the details, what was done about it, and then what the outcome was. The client modified the process to take into consideration mindset. They did this by tweaking some of the targeted selection questions in order to attempt to ascertain whether the candidate had a fixed or a growth mindset. I have seen some of the questions in books and so I am sure they were not unique to this particular client. Apart from the more conventional targeted selection questions, they also included questions such as:

- Tell me about a time when a project or action you did failed and your team knew all about it. What happened and what was the result?
- If you are successful in this application, what, if anything, do you need to learn to excel with our company over the coming years?
- Share with me an example of when you have been out of your depth in terms of an essential skill as a leader.

As you can see, the intent of these targeted selection style questions is to explore whether the person is thinking about what they have learned, how they learn, and that they still have learning to do. The answers their candidates gave helped in the selection of leaders who knew what they knew, knew what they didn't know, and had a mindset that tended more towards a growth mindset than a fixed one.

Has a High Level of Understanding and Curiosity about How Work Is Actually Done

Deeply understanding and being curious about how work is actually done has to do with how the real world works. It allows me to ask better questions of leaders and to challenge the status quo regarding procedural compliance expectations and how we think about incident investigations that have a human performance component. This means that we can learn not only after a workplace incident but also before a workplace incident, during normal day-to-day work.

In order to maintain a solid level of understanding and curiosity around Work-As-Done, Work-As-Normal, and Work-As-Written, we need to talk about these terms and be interested in them. We can establish some simple routines that revolve around talking about how work is actually done with our peers and with our teams. This could be in meetings, during pre-shift conversations, during field leadership interactions, or during casual conversations about how work is progressing.

During incident investigations and learning studies, the details of which are covered in detail in Chapter 4, we will talk extensively about Work-As-Done, Work-As-Normal, and Work-As-Written as we create the timeline with its elements of interest and as we try to understand the incident. These discussions and explorations of how work is actually done will help decide whether to focus the study on the differences between Work-As-Done and Work-As-Normal or the differences between Work-As-Normal and Work-As-Written.

Many organisations require leaders and peers of workers to spend some time out in the workplace interacting with workers as they do their work. These field leadership activities revolve around having conversations with workers on a day-to-day basis. As we watch people work and as we talk with them about their work, it is very easy to use the terms Work-As-Done, Work-As-Normal, and Work-As-Written as conversation fodder.

Here are some examples of work-related conversation topics that will help you be curious about how work is being done in the workplace:

■ How they are doing the task today.
■ How they have done the task before.
■ How the way the task is done may have changed or drifted over time.
■ How other team members and other teams normally do the task.
■ Whether the way they do the task exactly matches the 'procedure'.
■ Whether they have found and practiced a better way to get the job done than by following the 'procedure'.
■ How they may vary the way the task is done when things are a bit different in the workplace.
■ How any easier ways of doing the task are commonly done.
■ What their views are on the complexity, simplicity, and useability of the 'procedure'.
■ Whether they have seen other businesses doing the task in a better and maybe easier way.
■ Whether they feel the business has put up any barriers to them getting their work done.

As leaders, safety representatives, or safety professionals, we are often asked to review incident investigations. This is also a great place to show an interest in Work-As-Done, Work-As-Normal, and Work-As-Written. During an incident investigation review, we can establish whether the investigation team has really explored and understood what gaps there may have been between Work-As-Done and Work-As-Normal and/or between Work-As-Normal and Work-As-Written. We can also explore whether they have determined what can be learned through understanding what is driving those gaps. During regular leadership team meetings, go through a workplace incident investigation or learning study report in detail, looking at its quality. Include a conversation about how the work was actually done compared to how we thought it was being done.

As we create or review procedures, guidelines, work instructions, or manuals, whether after an incident or as a result of a routine periodic review, we can explore whether we are setting people up for success. We can do this by exploring whether the Work-As-Written is explained in such a way that maximises the likelihood that Work-As-Done will align with it. We are looking for accurate 'procedures' that are easy to follow correctly and difficult to follow incorrectly. In other words, the most logical and sensible way to get the task done aligns with the requirements of the 'procedure'. This translates to Work-As-Done equalling Work-As-Written.

As an example of how the ideas of Work-As-Done, Work-As-Normal, and Work-As-Written can pass from use solely within the confines of workplace incident investigations and learning studies into the broader business and leadership vernacular, I will share a recent story with you. I was engaged in a project, part of which was the training of every Superintendent, Manager, General Manager, and above in a large multinational minerals resources company on how to use Work-As-Done, Work-As-Normal, and Work-As-Written in investigations. About two-thirds of the way through this two-year-long project, one of the senior leaders told me that he was starting to hear Work-As-Done and Work-As-Written being used in leadership conversations and that leaders were starting to use the language in their field leadership interactions as well. This had developed without my input or intention. Large parts of the business ended up incorporating Work-As-Done, Work-As-Normal, and Work-As-Written into their formal field leadership conversation processes, data capturing, and auditing systems. Although I have often said that I believed that the conversations we have before a workplace incident should be the same as the conversations we have after an incident, I was pleasantly surprised that the terms had migrated into the conventional day-to-day language of the business. The message this sent me was that they had 'got it' and recognised the value in embedding such ideas in how we create safe work just as much as how we seek to understand and learn when things have not quite gone as we had expected them to.

Understands Their Own and Others' Expectations

Understanding the expectations we have of ourselves, the expectations we have of others, and the expectations that others have of us promotes and encourages the following sorts of statements: 'I really know what the boss wants', 'My boss knows what I want from her', 'I understand what my crew mates are trying to do and what they want from each other and from me'.

Out of all the elements of *Essentials of Safety*, this one lends itself best to promotion and teaching by the establishment of routines and then sticking to them. There is no rocket science to this element. Simply get into the habit of talking about the expectations you have of those around you and asking them what their expectations are of you. This can be in meetings, when talking about meeting outcomes, during employee performance review processes, daily pre-shift meetings, and task allocation activities. There are also many other opportunities to talk about expectations in your day-to-day leadership activities. Setting expectations does not need to be formal. Simply including 'what I want is …' will usually be more than enough to share your expectations with others.

I saw a great example of this recently as I sat at the back of a shift start meeting at a mine site during a maintenance shut down. The shift boss/supervisor was handing out work to a small group of workers. He went through what tasks were to be completed that day for each work group and then said:

> When you leave here, you need to go get together into your work groups. Once you are at the job, talk through the exact details of the job, who else is around and complete your task-based-risk-assessment. If you have any, and I mean ANY problems with working out exactly how you will do your jobs, or have not got the right tools or whatever, call me. STOP and call me. Is that clear? Is everyone happy?

Once he got a 'Yeah' from some of the group, he went on to say: 'Alright, what do you want from me? What do you want me to do to help with the job? Is there anyone I need to talk with, hassle or annoy to help get you going?'

This was a classic example of a leader clearly setting out his expectations about getting started for the day and also doing what he felt he could to understand the crew's expectations of him – well, at least to ask if they had any. It was not an attempt to deeply understand their general expectations of him as a leader. The pre-shift meeting was not the forum for that conversation.

In another example, this time sharing with you is something that you should not do to show you are clear and consistent in your application of your expectations. I was facilitating a workplace safety incident investigation and the subject of procedural compliance came up in an interview with the operational manager accountable for the area where the incident occurred. She felt that everybody knew that she was very strong on procedural compliance as she always told the workers to always follow procedures and she said she came down strongly on those who violated site procedures and got hurt as a result. When I was talking a bit later with the Safety Manager, it turned out that many of the procedures were complex, not aligned with each other, difficult to understand and to follow, and the operation have been working for a couple of years on trying to work out how to fix them. When I was talking with some of the workers however, it became pretty clear, pretty quickly, that it was all words and that the real expectation was that 'I will tell you that procedural compliance must be absolute, but I know that you cannot follow a lot of them. Just don't get hurt'. This is clearly a case of a lack of clarity on the expectations of the leader as the rhetoric does not match the practice. The result is that people hear the 'behaviour' and not the words of the expectations. So, whichever way you choose to show that you know and understand your own and others' expectations, the most important thing you can do is to be consistent and clear. And then keep talking about them.

Outside of the normal conversations on a daily basis, a powerful way to seek to understand the expectations we have of ourselves, the expectations we have of others, and the expectations that others have of us is to run a series of workshops with your teams. Before the workshop, spend some time thinking about the first two: the expectations we have of ourselves, and the expectations we have of others. Include some soak time in your thinking and revisit your ideas a few times before the workshop. Over the course of the workshop, you should make sure you are clear in what you want from yourself and what you want from others.

Start the workshop by setting the context - talking about what expectations are and why having clarity about them is so important. Once this is done and once everyone understands why you are there, talk through the expectations you have of yourself and how you will seek to behave. Include the 'Why' of these expectations as well. You can then talk about the expectations you have of others, specifically those in the room with you. These could include expectations around:

■ Procedural compliance, and what you expect people to do when they can't follow the rules.
■ Speaking up – but only after you have said what the expectations you have of yourself are with respect to listening.
■ Getting involved in problem-solving – those who do the work are the experts after all.
■ Being open to new ideas.

- ■ Doing what you say you are going to do or tell me early.
- ■ Working with honesty and integrity.

Let the workshop participants talk about whether they think the expectations you have of them are okay and that they can live with them. If not, talk about them and negotiate.

Once the team has clarity on your expectations – of yourself and of them - the time has come for them to come up with their expectations of you. You may need to prompt them a bit in terms of what sort of expectations other teams have of their bosses but try to limit this and encourage the team to come up with their own. Some examples with which you may need to prompt the workshop participants could be around the following:

- ■ Being consistent in your communication – don't confuse us.
- ■ Giving recognition when it is deserved.
- ■ Giving us feedback on performance – not just feedback on bad performance.
- ■ Looking to understand and learn, rather than automatically blaming us when doing a workplace incident investigation.
- ■ Being open to new ideas.
- ■ Doing what you say you are going to do.
- ■ Working with honesty and integrity.
- ■ Having our backs – putting your people first.
- ■ Being honest.
- ■ Involving us in decisions and problem-solving.

At the end of the workshop, you should be in a position where your team knows the expectations you have of yourself, the expectations you have of them, and the expectations that they have of you. You can then have ongoing and regular conversations about how you are all going with respect to the expectations you all have for yourselves and of each other.

Understands the Limitations and Use of Situational Awareness

I seek to understand the limitations and use of situational awareness to be able to talk knowledgeably about how situational awareness can be a benefit as well as a risk in the workplace. This helps me to look at things differently and helps my peers and workers look at things differently. 'Look at things differently' in this sense means that having an outlook that recognises we cannot see everything around us, the mental model we have does not totally represent a 'true' reality, and that we need to be selective observers of our workplace as we perform work.

Practicing this is not always easy. The best way to get used to exploring situational awareness is to practice. We need to get our bodies and minds into a routine of being observant. Take a few seconds each time you enter a room, approach a work group, or enter an unfamiliar situation, and even when you enter a familiar situation. During those few seconds after you enter, look around. Look for things that might be important to you and the group. Look for things that may be a risk now or into the future and decide which things are important to keep an eye on in case things change.

When you are in your workplace with a team mate/peer or with someone who reports to you and you want to help them to also take a view of situational awareness, do exactly the same thing as you would do on your own, but do it as a conversation. Ask things like:

■ What is going on here?
■ What do we see here?
■ What is important in what we are looking at?
■ Why might what we see be important?
■ What are people doing?
■ What is interesting about the work people are doing?
■ What around the people doing the work is interesting?

These sorts of questions can raise our level of situational awareness as well as help develop a shared mental model of the work. You can also develop the habit of encouraging work teams to do the same as they enter their workplaces. It takes less than a minute to do this. It is best done out aloud, even if you are on your own. It can have similar effects as telegraphing deliberate action, which we talked about earlier.

When you are interacting directly with workers, take the opportunity to ask additional questions such as:

■ What is going on around you that doesn't directly impact your work, but could if things slightly changed?
■ Let's just have a look around. What do you see that is interesting?

I was in a mobile equipment workshop recently, visiting with the local supervisor. As we left his office and entered the workshop, I asked him to stop and look around for a minute or two before we went and had a chat with some of the mechanics. We talked about what he saw and what he thought about it. Directly in front of us was a 793 Caterpillar haul truck. If you are not familiar with one of these beasts, they are massive. The tyre alone is about 13 feet (just shy of 4 metres) high. Its total height is about 21 feet (6.5 metres) and has a gross weight of over 324 tonnes. There was a maintainer standing on top of the tyre in front of us. He was wearing a harness and a lanyard and it was connected to the truck at the level of his feet (3.6 metres above the ground). The supervisor did not raise it as being of interest and so I prompted him with: 'What are your thoughts about how this work is being done here in front of us?' He responded by: 'It looks okay to me. John has his harness on and is connected up, so it all looks good'. It was very clear to me that he did not see the situation in the same way that I did, so I asked him to talk about what needs to go right in this task instead of starting with what could go wrong. Again, he suggested that the maintainer needed to be connected to an anchorage point at all times. I then asked what could go wrong, what controls are in place, and how those controls work. He was getting a bit frustrated until he finally saw the situation as I did and realised that the harness was never going to stop the maintenance worker from bouncing on the concrete if he slipped off the tyre. He then stopped the job, got the guy down, and worked out how the task could be done a bit more safely. We retired back to the office and chatted about the different ways we see things. I took the opportunity to talk about situational awareness and how we can choose to look at things however we wish to. We talked about it being more difficult to see things in a very familiar environment and when doing routine tasks. I have been out with the same supervisor a couple of times since and he told me that he always stops when he leaves his office and spends 20-30 seconds to look at the workshop by using 'a Longy's eyes look'. Doing practices and routines like this interaction can greatly assist peers and workers to look at things differently, to see things differently, and to seek to understand the situation in which they are working. This includes thinking about what impact the specific work situation may have on the worker or on those around them.

Listens Generously

For me, to listen generously is to send strong messages of interest, focus, attention, and care, so that people know I am listening to them. Generous listening also sends the message that what others are saying is important. Listening helps me be a better leader as I learn what people have to say and how their mental models match or do not match mine, what their views on a topic are, and what their answers to problems are. My belief is that listening is the most important skill a leader can possess. Listening is a skill that I constantly try to practice. Listening can be learned and practiced in nearly every situation.

Generous listening lends itself to establishing and practicing routines for every level of leader and worker. One of the most effective practices to improve your listening is to think about how you are listening – I have heard this called meta-listening. Thinking about your listening can be easily done after a conversation. Making it a routine after each conversation will result in dramatic increases in your listening skills in a short period of time. Get into the habit of conducting a conversation post-mortem, examining how well you listened, what worked in your listening and what didn't work, and what you could do differently the next time. Especially focus on whether you led the conversation, put words in the other person's mouth, answered your own questions, or did too much talking at the expense of listening. Keep tabs on the outcomes of the post-mortems and try to improve on subsequent conversations.

Another simple practice is to count. Whenever you have an opinion or have something to say, hold back and count to three to see if the other person has something else to say first. When you do speak, try to build your thoughts onto what the other person had just said. This is a skill. Build on what others say sends a strong 'I am listening' message. You can observe this behaviour with expert interviewers on television. Try to emulate them. They build their question on the shoulders of the previous answers. To see what this looks like, have a look at some of the Michael Parkinson interviews on YouTube.

The other simple routine is analysing listening during meetings. Establish a listening coach and allow them a few minutes at the end of the meeting to give the other participants in the meeting feedback on the generosity of the meeting participants' listening. I covered this in Chapter 1, but it is worth reiterating here as it is such a powerful method.

You can get someone to be a listening coach in just about any meeting or activity. It can work when you run pre-start meetings at the start of a work shift, when undertaking coached field leadership interaction activities, or even during training sessions. The listening coach role is in addition to being part of the meeting or activity. It is in addition to, or it can be combined with other common meeting process roles such as the process checker or timekeeper.

The role of a listening coach is to listen to the people and to give feedback on how effectively the participants, leader, or presenter listened during the meeting or activity. A listening coach looks for a number of things:

■ The degree to which people are building their thoughts and words on what the previous speaker ended their talking with, or whether they go off on a tangent or immediately introduce an unrelated concept or topic.

- The amount of time people give when somebody wants to say something. If or how much they interrupt either verbally or by their body language.
- The body language of the listener, including eye contact, body position, and focus.
- The number and intent of interruptions.
- Generally, whether the team members are listening generously or listening stingily.

Another opportunity to develop your own and others' listening skills is during performance appraisal reviews with your people. Monitor both your own and the other person's listening style as you go through the process. You can also include generous listening into someone's development plan and, of course, do your own conversation post-mortem after the discussions have wrapped up.

There are some great little practical exercises to help people improve their generous listening skills. One I like and have used a lot – mainly during post-incident interview training – is to break the participants into pairs. Get one member to tell a story about something they love doing outside of work. The other person is to listen as hard as they can using body language along with affirmation etc. to show they are listening. Let this go on for maybe 5 minutes. Switch it up and move people into different pairs. Get the person who told the story the first time to repeat it to the new participant. This time the second person is instructed to actively *not* listen. Do not tell the story tellers in either case what the listeners are doing. Just tell them to tell their story twice. Get the *not listeners* to look around the room, check their watches and phones, yawn and otherwise not listen to the story. After the stories are told, get the story tellers to talk to the group about their experiences. This activity is very enlightening to show people what it is like to be listened to generously and what it feels like not to be listened to at all. This will hopefully jolt them into some generous listening practices.

Always remember that thinking about our listening improves our ability to listen. This is why doing conversation post-mortems are so important to help you improve your generous listening skills.

Plans Work Using Risk Intelligence

Why do I seek to plan tasks using tools such as the Efficiency-Thoroughness Trade-Off (ETTO), Risk Intelligence, and a suitable wariness for the effectiveness of controls? I do this by trying to keep a focus on these concepts during minute-by-minute and day-to-day planning. This ensures that I spend enough time thinking before doing the planning as well as before doing the work.

Exploring what needs to go right, what could go wrong, and how efficient or thorough I need to be so that things, on the whole, go right, rather than go wrong.

It is a simple matter of getting into the practice of exploring each of these ideas as you go about planning and then doing work.

There are some simple questions that you can ask yourself as you start to think about work:

- How thorough do I need to be here? How efficient can I be? If I get the balance wrong and am too thorough, I may be wasting a heap of time for no good reason. If I am too efficient, then maybe safety, quality, or production mistakes will be made. It is all about getting the right balance.
- Do I know what I need to know here? Do I have the knowledge to be able to make good decisions or do I need to talk to someone else before I get too far down the planning route?
- How confident am I with respect to the controls actually working as I think they should?

These ideas can also be easily woven into conversations with workers and leaders during field leadership interactions. By asking very similar questions as you ask yourself, you can engage in a conversation, the outcome of which will be a shared mental model of the work and a greater understanding of the way the work is actually being done.

I was talking with a frontline supervisor in an underground mine and he told me how he manages this aspect of the *Essentials of Safety*. He does a lot of driving and a lot of juggling of people and machines during the day. It is always about what to do now, what to do later, what problems suddenly rear their heads and need solving, what changes people's wants for the shift, which bit of equipment needs to be moved for this or that reason, which visitor needs to be shown around, what field leadership activities need to be done, or what tools need to be picked up and dropped off somewhere in the mine. It is a never-ending and tiring battle. After talking about the Efficiency-Thoroughness Trade-Off, Risk Intelligence, and a suitable wariness for the effectiveness of controls with him, he put in place a little routine that he said works really well. After the conversations with the previous shift supervisor during handover, and before addressing the troops at the pre-shift meeting, he jots down a few notes on the back of a little card that he refers to during the day. He puts down the main jobs for the day and what is supposed to be achieved for each of them. He adds the things that he wants to achieve. For example, he may add FLI, which is his shorthand for 'field leadership interaction'. On the front the following is already printed:

Interruption prompts: Answer these questions each time a new request or issue comes up:

What is the ETTO balance?
What is the priority? Urgent – Not Urgent
Is it a Need or a Want?
Will it result in more or less control effectiveness?

He uses this card each time he gets a request to do something. He finds this helps him prioritise his planning. He loves it. I encouraged him to share it with other supervisors.

In summary, it pays to maintain the balance during your planning and your conversations. You cannot have both thoroughness and efficiency as your masters at the same time. It is always a trade-off. You need a balanced view with respect to chronic unease versus lackadaisical carelessness, and you also need to maintain a balance between absolute confidence and a wariness of the effectiveness of controls. Getting the balance right requires leadership, conversations, and having time to think as you set up the work. The most important routine here is to simply pause and think before you do any planning. This applies as you go about your normal work, regardless of whether it is formal, or last-minute, day-to-day planning.

Controls Risk

To seek to control risks is to maintain control over the workplace – to be aware of what is going on around me and preserving options if something does not go according to plan. To control risks is to assure myself and to show others that I can maintain the balances in the workplace. To be situationally aware, have a clear mental model, plan effectively, understand the hazards and risks, talk thorough my actions by telegraphing deliberate action, undertake CCVs (Critical Control Verifications), review normal work using Learning from Normal Work Reviews and being mindful and authentic, so that those around me also pick up these practices for the benefit of us all.

Although 'Controls Risk' is such a broad topic, there are a few simple routines and practices that keep it front of mind and are effective in checking both ourselves and others as we control risks whilst doing work.

Some simple routines can start very early on in the processes associated with creating and doing work and do not all need to be done when in the field watching real work being done.

Things to do before the work has started:

- As you develop work instructions and/or procedures, routinely ask about any trigger steps and critical steps and if they can be built into the procedure.
- Each time you create work environments and safety systems, attempt to create an environment where people can fail safely. Because fail they will.
- Get into the habit of undertaking Learning from Normal Work reviews.
- As you undertake your routine reviews of material and/or fatal risk databases and significant risk bow ties, check that the critical controls are clear, make sense for those who have to implement them, and are few in number.
- Practice telegraphing deliberate action and encourage others to do the same.
- Build as many of the *Essentials of Safety* elements into the creation of work.

Whenever you are out and about having field leadership conversations, explore some of the topics below:

- Ask about trigger and critical steps as the work is actually being done. See if the workers know what they are and what they need to do with them. Explore how we make sure the trigger steps and critical steps go as planned.
- Check if the individual/team members have escape options in case something goes wrong.
- Talk about telegraphing deliberate action and encourage people to adopt the practice.
- Regularly do Critical Control Verification (CCV) activities. (The purpose of the CCV is to verify that the critical controls from fatal and material risks are in place and effective so that we can provide ourselves with a level of assurance that each time the activity is undertaken all critical controls are implemented as designed.)
- Get into the habit of asking what other tools and equipment, procedures and systems could be used that would make life easier for ourselves and those doing the work.

- Check out if the team has a plan to make sure everything continues to go right.
- Routinely ask people why they are doing what they are doing and what is going on around them.
- Engage the team in a conversation about what is in their line of fire and what they are doing or are planning to do about it.
- See if the team or individuals are aware of what devices or elements provide them with alarms and warnings when danger is imminent.
- Explore whether there are any anomalies or changes in the workplace that do not appear to have been there before. Check if they see anything that stands out as different to what was expected.
- Ask about their plan if something does go wrong.
- Ask 'What if?'
- Have they thought about what usually goes right here, but may not go right today?
- Talk about what they think will happen if they do not control the risks.

As I was writing this chapter, I saw a wonderful example of someone using telegraphing deliberate action without realising what they were doing - on one of my many Uber trips to and from the Perth airport. I observed that once the driver put my address into his phone, he spoke out aloud the turns before he did them. He did this all the way home, just quietly – almost to himself, but at a level still audible enough for me to hear him. We were going along the freeway and he said 'Taking the Thomas street exit' some 200-300 metres before the exit, and then he took it. A bit later he said 'Left on Olive'. I spoke up then and said 'Actually it is better going up Axon'. He went up Axon street. We then had a chat about the idea of telegraphing deliberate action and how it can minimise mistakes. He was very happy that he was doing something that made sense.

Applies a Non-Directive Coaching Style to Interactions

I use a non-directive coaching style as often as I can so that I can really help others be the best they can be, help them solve their own problems, and see examples of how using a coaching style may be something that helps with their leadership.

Teaching people how to be great coaches is quite simple. It is something you can do both from within and from outside formal training sessions. You can teach yourself and others to be great coaches by simply doing it a lot. It becomes a practice and routine that just flows.

Before someone has achieved learning with respect to coaching, they need to understand the 'Why' of coaching and why using a coaching style will help them in their leadership as well as making life easier.

To order to assist both yourself and others see the 'Why?' of coaching, ask questions like:

- Why would I be interested in becoming great at coaching?
- Why would I be interested in getting better at helping others get really good at leadership, field leadership conversations, and safety interactions generally?
- Why would I be interested in getting really good at helping people get better at coaching?

I hope you come to the conclusion that coaching, and using a coaching style in your leadership, makes life a lot simpler and more effective than otherwise it might be.

In terms of self-education and routines to advance your coaching skills, get yourself into the routine of listening and asking more than telling. Whenever someone asks you a question, stop and think about whether just answering it is the most valuable thing you could do, or would adopting a coaching style be better? A coaching response may not seem to be the best thing for you at the time necessarily, but it may well be the best thing for the person asking the question, and the best thing for you in the long term. Through thoughtful questions and processes, think in what way the player may be guided through your coaching to come up with the answer to their questions themselves. People coming up with the answers to their own questions yields a more powerful and sustainable learning than you simply telling them the answer. This is where one of the powers of coaching manifests.

Another routine you can practice is to practice using the GROW model on yourself. At least the R, O, and W bit of it. When you have a problem to solve, firstly self-explore exactly what the current Reality is. This will of course be your perception of the Reality, but that is perfectly okay in this case. Try to look at your Reality from different angles and perspectives. Once you have clarified to yourself what the current situation actually is, explore what Options you have and then come up with what you Will do about it.

And remember that Coaching is all about you, as a leader, helping those around you be the best they can be by listening, observing, asking questions, challenging, and caring, all whilst being authentic to yourself as a leader and as a person.

Has a Resilient Performance Approach to Systems Development

To adopt a resilient approach with respect to systems is to have a focus when undertaking systems development, along with checking those systems' effectiveness in the field so that they are resilient to the vagaries of real work.

There are two areas where routine leadership practices demonstrate applying a Resilient Performance approach during system development that stand out for me. One is in the development of safe systems of work and the other is during field leadership conversations with workers who aim to verify the implementation of those systems. Looking at the development of safe systems of work initially, we can get into the routine of asking what a system will look like, who the audience is, what the details will be. We can also ask ourselves, and any others involved in the creation of the system, some of the following sorts of questions:

- Is the detail within the system going to encourage or promote those using the system getting things right? Or is it going to encourage or promote getting things wrong?
- Does the system/procedure or whatever it is tell users what to do when things start moving away from going right?
- Does it offer advice on what to look for or what to monitor that needs to be in place to ensure things go right?
- Does it contain the lessons learnt from previous successes and failures?
- Does it point to common elements that the users may expect to develop into the future of the task or system they are using?

Also, in terms of system development, get into the habit of exploring the 'maximise' and 'minimise' elements of the Safety Oscillation Model described in Chapter 1. We do this so that the system as it currently exists is supporting the 'maximise' and 'minimise' elements of the model rather than actively pushing in the wrong direction. Until we look at our systems through the eyes of a model such as the Safety Oscillation Model, it is often difficult to see what the impact of a new system or procedure, or a change to an existing one, will have on the overall direction of safety.

For example:

- When there is an increase in 'production pressure', are our current systems and processes actively working to minimise the negative impact on the 'effort on safety'?

- When there is pressure to reduce the 'business budget', do the systems protect the 'budget for safety'?
- When 'production pressure' is reduced just after year end, does the system work to minimise 'complacency' and maximise the 'effort on safety' through tweaks to the systems that assist in the creation of safe work?
- When things are going really well, as in an increase in 'safety outcomes', do our systems unwittingly drive the business to reduce its 'effort on safety'?

The other area where routine leadership practices can adopt a resilience performance approach lies with the field leadership conversation process. Here you should develop routines in your conversations to explore how resilient the workplace is as you chat with the workers. You can make a difference simply by being authentically interested in understanding the mental models of those doing work. Have they thought about what they need to make sure they have a bit of a plan if things start to deviate away from going right? Do they know what to keep an eye on? This is related to their level of situational awareness of course. Do they understand how the procedure they are following, or not following, works to help them do safe work? Do they know of, and have learned from previous successes and failures? And do they share their mental models of the work with each other?

In summary, be constantly vigilant as you explore, review, tweak, or verify systems. Aim to apply a resilience performance approach anytime you get anywhere near one of your systems. Especially remember that it is often the interrelationships between systems that cause problems, not just a particular system on its own.

Adopts an Authentic Leadership Approach When Leading Others

To lead authentically gives me an opportunity to share ideas, concepts, and practicalities in leadership with as many people as will listen, so that people start to think differently and positively about the why, what, and how of the things they do in leadership. You may notice from Chapter 1 that this is virtually a repeat of my own 'Why'. This is because I truly believe that authenticity in leadership is *the* most important element in this book. Through relationships, authentic leadership at all levels drives absolutely everything else in safety.

In many ways, all of the elements in *Essentials of Safety* are about leadership, and so practicing all of the elements described in this and other chapters represent the routines you need to adopt for authentic leadership. The only additional one is to constantly check that you are being authentic. Ask yourself all the time whether what you are thinking, saying, and doing is really from your true self. That little voice in your head does not lie, so this is a very simple thing to do. If you find yourself justifying your view, then you are probably not being authentic.

Some routines that will help you keep an eye on your authenticity include:

- Routinely asking peers about how you come across as a leader.
- Telling stories about your successes and failures in your leadership journey.
- Asking your people to tell you what is really going on in the business and then listen to it.
- Reviewing your week for consistency in leadership conversations and decisions.
- Weekly post-mortem on leadership activities – what worked, what didn't work, and in both cases exploring why.
- Ask yourself on a monthly basis:
  - What does being authentic mean to me in my life?
  - How can I be more supportive to others?
  - What support do I need from others?
  - Have my leadership activities been in alignment with my 'Why?'
  - Is my 'Why' still valid?

At the end of the day, as Goffee and Jones say in *Why Should Anyone Be Led by You?*, authentic leadership is all about 'Be yourself – More – with Skill'. In other words, to be a leader, you must be yourself, only more so.

Summary

Leadership practices and routines will need to be unique to your leadership style. The topics, practices, and routines above are designed to pique your interest in developing ones that work for you. In the same way that no one leader's particular style will work for everyone, there is nothing worse than all the leaders in a business parroting the senior manager or the flavour of the month leader from Forbes or HBR. The routines and practices around authentic leadership need to be owned by you, the leader. Think about which elements of your leadership style make you the unique leader you are and develop appropriate practices and routines that develop and hone those elements.

Chapter 3

Barriers and Their Remedies

> Barriers are in our minds. Remedies are in our words that become our behaviours.
>
> **Ian Long**

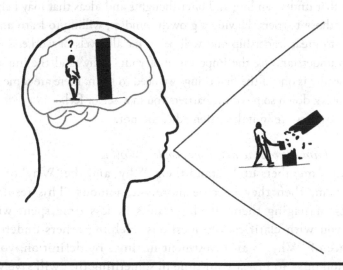

Figure 3.1 Barriers are in our minds. Remedies are in our words that become our behaviours.

DOI: 10.1201/9781003181620-4

Introduction

The purpose of this chapter is to talk about the challenges and barriers that may arise as you start to adopt the essentials of safety and then to talk about remedies to those barriers. Barriers may arise not only within one particular *Essentials of Safety* element but can manifest across more than one of them at the same time. This is why you will see some of the remedies for barriers looking and sounding similar across the elements. When you are thinking about and then designing your own remedies, bear this in mind so that the result is a simple approach and not a hotchpotch of ideas and behaviours. I have attempted to make the barriers and remedies as specific as I can. You may well need to tweak the ideas to match your particular barrier and remedy to suit your needs.

Understands Their 'Why'

If we understand what we do and we understand how to do what we do but do not really know or truly understand why we do what we do, there can be trouble. A big part of our role as leaders, regardless of our level in the hierarchy of the business, is to help ourselves and others be the best we can be. Understanding our 'Why' is a major part of this. If you do not understand why your 'Why' is important, nor are you particularly interested in learning why your 'Why' is important, then you are in the right place in this book. Because here is where we explore what some barriers may be to that understanding and offer thoughts and ideas that may help remove, or remedy, those barriers. Having a growth mindset, willing to learn and grow is imperative to great leadership and well-being for all levels of workers. Removing barriers to understanding the importance of your 'Why' and the importance of 'Why' generally is one of the first things we need to learn. Here are some ideas that may help break down some of the barriers you may see or feel.

Note: Barriers are in italics. Remedies are not.

As a leader, I am too busy to ask what people's 'Why' is:
If your team members understand their 'Why' and the 'Why' of the tasks you set them, then they become more autonomous. This results in less time micromanaging them. It also results in less time spent when they come to you with clarification questions. Helping others understand the importance of 'Why' is an investment in time. By definition, you should never be too busy to invest your time in something that will save you time. The remedy therefore is to treat the time asking about people's 'Why' as an investment in freeing up your time.

I don't want to ask about other people's 'Why' – It is too personal and can also be a difficult conversation to have:
This barrier usually presents when the leader does not truly get their own 'Why' – why they are a leader – why they do what they do. The reason I say this is that if a leader really understands their 'Why', then they will do what needs to be done, including things that may feel uncomfortable and are sometimes difficult, knowing that they are important. This idea ties closely in with our attitude. We can always choose how we react to various tasks we need to undertake. We can if we wish choose to consider the task of helping people explore their 'Why' as a task that will help them immensely as they strive to be as best they can be. This helps us overcome any difficulties in the conversations because we know the value of that conversation to ourselves and those we lead.

I don't know my own 'Why'. How can I ask about someone else's?
The simple remedy here is to spend some time, with some help if needed, to explore what your 'Why' is.

It is none of my business what other people's 'Why' is:
The solution here is to spend some more time understanding what your role as a leader is. This is talked about in Chapter 1. In a nutshell, I believe that one of the significant roles you have as a leader of people is to help others be the best they can be. A significant way you can assist them in this is to help them understand their 'Why' and to have conversations with them during the day-to-day about the 'Why' of the various tasks and activities that they are required to do. A worker at any level, including all levels of leadership, who understand their role – why they do what they do – will be far more effective, happy, and autonomous than one who does not know why they do what they do.

I have never thought about asking about a person's 'Why':
If you understand the importance of understanding your own 'Why', then this remedy will be obvious and maybe somewhat superfluous. The solution is to make *Understands their 'Why'* part of your leadership conversations, routines, and annual performance appraisal processes within your business. Talk about your own 'Why' and have conversations with people about their 'Why'.

I have never thought about why I do what I do. I have never stopped to think about what my 'Why' is:
Hopefully by the time you are reading this chapter, you are now aware that thinking about what your 'Why' is can be very valuable. Recall that we know

what we do. We know how we do what we do. But not many people actually know why they do what they do. Once you know why you do what you do, the workplace and the world look different and usually much more meaningful and exciting. The remedy is as per a few above. Spend some time and explore your 'Why'.

I have never been asked by my leader or anyone else what my 'Why' is:
That is okay. Just because you have not been asked what your 'Why' is does not mean that you should not do something to work out what it is. The simple remedy here is the same as a previously described barrier and remedy: to spend some time exploring what your 'Why' is. NIKE – Just Do It.

If I ask people why they do what they do, they often jump quickly into 'Because it is my job', 'Money', 'Because this is what the boss wants me to do' and do not seem to want to go deeper and as their leader, I have not got the skills to help them:
It is worth spending time, once you have worked out your own 'Why' as a leader, to spend time together as a leadership team to learn how to help others get theirs. It is not a difficult process but it does require rigour and a good dose of authentic leadership.

'Why' is not talked about in one-on-ones with my boss:
Ask your boss to include *Understands their 'Why'* as a part of their leadership team conversations, routines, annual performance appraisal processes, etc. within your business. If this does not work, it is still really beneficial for you as an individual to spend the time and interest in exploring for yourself why you do what you do, inside and outside of work.

I do not talk about my 'Why' or my people's 'Why' during my one-on-ones, nor during their annual performance conversations – it is not part of the HR format we need to follow:
As above, simply make *Understands their 'Why'* part of leadership team conversations, routines, annual performance appraisal processes, etc. within your business. Make changes to the Human Resources and management systems that will support and facilitate having 'Why' as a routine part of the processes that drive performance reviews and other people's related activities.

'Why' is not something we talk about during training – 'Why are we doing this training? – What is in it for me? What is the context?' – This stuff is not talked about:

This is one of the simpler remedies and is very easily implemented. Modify every training process to start with the 'Why' of the training session. Every training and development activity has context and purpose. Expand these; formalise them into the 'Why' of the session. As those in training workshops get to understand the 'Why', you will find that this also improves the quality of the learning.

To get to my real 'Why' I would need to think about it a lot. Not something I have time for at work, nor the inclination:
As a leader, understanding your 'Why' is an investment in time. It is not something that is done and forgotten about. Yes, you will need to take some time and yes, you will need to think deeply and hard about it. But it is worth it. People at all levels, including all levels of leadership, who understand their role and why they do what they do will be far more effective, happy, and autonomous than ones who do not know why they do what they do.

My personal purpose and 'Why' has got nothing to do with my boss or my work:
Even if you feel this, it is still very important that you know why you do what you do – your 'Why'. If you are crystal clear on this and it aligns with getting the job done, then not sharing it with your boss is not a problem. You are the one person who will directly benefit from understanding your 'Why'. Having said that, it can make work flow more easily around you if others are also aware of what drives you.

I have been asked what I am doing and how I am doing it but I have never been asked why I am doing it:
Talk with your boss about making *Understands their 'Why'* part of the leadership team conversations, routines, annual performance appraisal processes, etc. within your business.

I do not see why knowing my 'Why' is of any use to me:
It is sometimes a bit of an unknown unknown. It is often difficult to determine the value of a process if you have not been through the process. My strong suggestion is to learn what drives you – your 'Why' – and then make the comment as to whether or not knowing your 'Why' is of any use to you.

So, in summary, the simplest way to get people to understand their 'Why' is to talk about it often. Include the 'Why' whenever there is a training workshop, regardless of the topic. Talk about your own 'Why' during development and performance appraisal activities. Bring it up when you are talking with peers. Make the discussion of 'Why' a part of what we do around here. This remedy also goes a long way to telling your people that it is perfectly okay to talk about what drives you, why you do what you do, and that it is a normal part of work life.

Chooses and Displays Their Attitude

The concept of a barrier to people not choosing and displaying their attitude could be construed as a bit of a misnomer. By this, I mean that people will usually display the attitude they have chosen and so the concept of a barrier that prevents them from showing their attitude is a bit weird. The issue here is when you, or individuals or groups of individuals in your team, are showing that their attitude doesn't auger well for a happy and productive workplace. So, the barriers tend to be barriers to having a team full of people choosing a positive attitude as compared to barriers to people choosing and displaying their attitude *per se*.

We often, or at least usually, expect the behaviour of a person to be consistent with the attitudes that they hold. This is not always true and this is where cognitive dissonance gets a guernsey. For example, people smoking whilst knowing that smoking causes lung cancer and heart disease. So, the endeavour here is to remedy those situations where behaviour seems to indicate an attitude that we may want to turn to a 'better' one.

I do not understand what you mean by 'attitude'.
When we do work, we do not always have a choice about the things we do in our work. Some tasks we simply have to do as a part of our job. We do not have to be ecstatic about doing the tasks, but we do need to choose how we react to the tasks. We can decide how we view those tasks. This is what choosing and displaying our attitude is all about. We get to choose how we react to what we need to do.

My team has a can-do attitude. What more do you want?
Having someone with a can-do attitude usually means that you approve of them because they are confident and willing to deal with problems or new tasks, rather than complaining or giving up. This is okay as far as it goes. What we are also striving for is that people are aware of the attitude they are holding at a particular point of time. Talk with your people about what attitude they seem to be displaying and check in to see if it matches what they are feeling or is a bit of a put-on.

Some people have different attitudes to different tasks. They like some tasks and so seem to enjoy doing them. Other tasks they do not like, and they make it known …:

The secret here is to educate your people so that they are aware of the impact their attitude can have on themselves and their workmates in terms of happiness, well-being, and productivity. In many ways this is tied into their 'Why'. If people have a clear sense of why they do the jobs they do, warts and all, they will tend not to display negative attitudes about them as much. Also, as they say, 'awareness is curative', so try to set up processes whereby it is safe to talk about the impact individual's attitude is having on others.

As a leader I need to always appear to be in control. I need to not show weakness and I always appear serious:
'Behaving like a great leader until you become a great leader' is not bad advice. However, it must be authentic. Always appearing serious does not usually come across as truly authentic leadership. Showing some weakness is a great leadership trait and one that should be encouraged when you come across this issue.

I just have to do the work. How I feel about it is not relevant:
If you have people with this attitude, you need to help them understand the impact it can have on their well-being and that of their peers at work. This can be done through training or more effectively through on-going conversations and interactions.

How can I have an opinion about a piece of work when I do not even know why I am doing it?
This clearly points straight back to the barriers discussed in *Understands their 'Why'*. It is of course entirely up to an individual as to the attitude they wish to have and display. It is up to the rest of us, whether peer or leader, to influence that through relationship and conversation. When somebody is showing up with a black attitude, we can challenge them to be aware of it and hopefully to change it. We can also front-end-load this element during the employee selection process – choose employees who, when given a hypothetical scenario, choose a positive attitude to handle it with rather than a negative one.

Adopts a Growth Mindset – including a Learning Mindset

As both a growth mindset and a fixed mindset are by definition mindsets, they are created and manifest within the mind of the individual. This means they have an enormous impact on what a person says and does. There is a Buddhist idea that what you think leads to what you say which leads to what you do – your behaviour. So keeping an eye on what we are thinking is always a good idea as this will impact what we say and do. How does this relate to having a growth mindset? The reason the Buddhist idea is relevant is that you can think your way from a limiting fixed mindset to an all-expansive growth mindset.

In the case of remedies to barriers impacting *having a growth mindset*, they are all around helping people move from a fixed mindset to a growth mindset. Consequentially, I have lumped all of the barriers together into one list and then drawn a common set of ways of thinking and doing that helps people move from a fixed mindset to a growth mindset. The barriers are these:

- *A lot of our workers say that they do not understand what you mean by a growth mindset or a fixed mindset.*
- *I have done this job for years. I am what I am – I could never learn to juggle.*
- *I have a fixed level of intelligence – I haven't got the intelligence to be a doctor.*
- *I have reached my peak in the brain race.*
- *I cannot change who I am.*
- *A leopard cannot change its spots.*
- *You cannot teach an old dog new tricks.*
- *This is how I am. This is how I have always been and this is how I always will be.*
- *I have not got time to read books and learn other stuff. I have better things to do with my time.*
- *Just tell me what to do.*
- *I am already good at my job.*
- *I do not want to get a promotion, thank you very much.*
- *I am happy just being me.*
- *No fancy pants university and managers job for me, thanks anyway.*
- *I learn nothing when someone gets up in a training session and tries to teach me by reading PowerPoint slides and asking me stupid questions. I just turn it off.*
- *I don't like being 'stretched'.*
- *Why does everything have to be about learning?*
- *Why can't we just fix what is broken and be done with it?*
- *I go and do field leadership activities to verify that our people are doing the right thing, not for me to learn. I already know how they should be doing the work.*
- *Others need to learn how to behave according to my expectations.*

These are classic examples of people who profess, at least in their words, that they already have the right mindset for the work they do. You will notice that this applies not only to followers but to leaders as well. This may or may not reflect that they have a growth mindset or a fixed mindset and so it is important not to make assumptions here. The trick is to start a journey of moving the individual from wherever they are at present to a growth mindset. There are a few simple ways to stretch them over time to think differently.

The secret to moving people from a fixed mindset to a growth mindset is in helping them understand the 'Why' of a growth mindset. This may well – and probably should – start well before you employ them. Start exploring people's ideas and thoughts about their mindsets during the interview stages of employment in order to set up for success later on. Choose employees who, when given a hypothetical challenge scenario, display a growth mindset when thinking about how to handle it. Then, during the early induction process start them thinking about their mindset in detail, setting challenges, and expanding their thinking. Every employee can benefit from a growth mindset.

Through annual performance appraisal processes, individual development plans, and other means, help people to learn that, just like their attitude, they *can* choose their mindset.

Teach them that they are not perfect just now and neither are you. We can all learn and grow – if we want to. Show in your leadership style, language, and behaviour that you are not perfect, that you embrace failure and mistakes as ways that you learn.

Link the way they view challenges to their attitude. We can choose how we react to challenges and helping people view challenges as opportunities can really help here. Allow them to fail and see each new challenge as an adventure.

Again, back to an earlier element, help them work on their 'Why'. People who have a clear purpose in their approach will tend to move towards a growth mindset more easily as they clearly know why they need to do what they need to do. Being authentic to themselves and their 'Why' will also help develop a growth mindset.

Encourage people to listen carefully to the words and thoughts that form in their minds as well as the words that come out of their mouths. This again ties closely in with choosing and displaying their attitude. When you talk about development plans, focus more on building people's strengths than 'fixing' their weaknesses. Teach people how to give, and more importantly how to take, feedback. Accepting feedback as a gift, regardless of how forthright it is, is a great skill to have and can be taught. I am reminded of a senior leader friend of mine who would always preface giving feedback to one of his team with 'Feedback is like going into a department store and trying on a shirt. You take it off the rack and try it on. If you like, it, you keep it. If not, then you put it back on the rack'. Any form of criticism can also be viewed as a gift, if you choose to see it that way.

Encouraging people to use the phrase 'not yet' instead of 'no' when talking about the skills they do not have. 'I have not yet become a great coach, but I am on the journey' or 'I am not a great coach, yet'. Both of these are much better than saying 'I am not a great coach'.

Having a growth mindset is such an important attribute that you must spend time cultivating it in yourself and in your team.

Has a High Level of Understanding and Curiosity about How Work Is Actually Done

I do not understand what you mean by Work-As-Done, Work-As-Normal, and Work-As-Written:
Start to use the language of Work-As-Done, Work-As-Normal. and Work-As-Written when you talk about incident investigations and learning studies. Be curious about each element as it relates to the incident being understood and learnt from. Use the language when you go into your 'field' and talk with the real people who are doing the work about what they are doing, whether that is how they and others normally do the work and what they think about the work instructions and procedures. Start to explore the differences between how the leaders think the work is being done and the way it is actually being done.

I expect my people to do work as per Work-As-Written:
The advice here is simple. Build relationships with your teams, talk with them, watch them work, and start to get a handle on how the work is actually done out there, on a day-to-day basis where real people control risks by doing work. There are very few businesses and industries where Work-As-Written equals Work-As-Done. The need is to start to challenge your thinking and to be honest about your expectations and reality.

Why would people not just follow the rules?
A great little experiment to help here is to pick a task that one of your team does and pull together *all* of the rules, procedures, work instructions, directions, guidelines, and standards that apply to that task. Lay them out on a large table and go through them in fine detail. See if it is possible to actually follow them.

In the vast majority of cases, you will see that it is not possible to follow all of the rules all of the time. This should trigger some serious curiosity about Work-As-Done, Work-As-Normal, and Work-As-Written.

One of our company's safe work behaviours is that our people will always follow procedures, therefore Work-As-Done must equal Work-As-Written or someone has violated a sacred rule on site (the safe work behaviours) and this is what will cause accidents:
See the remedy directly above. This is not the real world and such overarching rules are not conducive to trust, authenticity, and safety.

During incident investigations, we only look at the differences between Work-As-Done and Work-As-Written:
This is a very common approach across many industries and honestly does nothing to help create a culture that is in any way 'just'. The advantage of extending the view of looking only at Work-As-Done and comparing it with Work-As-Written is to attempt to explore when you see a difference between Work-As-Done and Work-As-Written, whether that Work-As-Done that you are looking at represents Work-As-Normal or not. In other words, is the way the work was being done at the time of the incident the way that everybody else was doing the task? If this is the case, then the Work-As-Done had nothing to do with the individual involved and represents a broader issue that needs to be understood and learnt from.

After an incident, the General Manager always asks 'Who is responsible for this incident?'
The answer to this question is, of course, 'The General Manager is responsible for this incident'. This does not usually help remedy this barrier much though. We need to help the General Manager change his/her question to 'What is responsible?' or 'What can we learn from this incident?' This can be achieved by helping them understand that the purpose of an investigation is to learn by exploring any differences between Work-As-Done, Work-As-Normal, and Work-As-Written. The purpose of an incident investigation is to understand and learn, not to blame.

As an operator, I know how the real world works, and it is not always the same as Work-As-Written:
Absolutely. As a remedy to hearing this barrier, we need to share it with leaders. It is not really a barrier at all. The lack of understanding or belief of it by leaders is the true barrier here. This truth needs to be understood by all levels of the

organisation, not just this doing the work. Honesty, openness in conversation and trust play a huge role here. Leaders and those doing the work need to communicate openly about how the world really works. This takes time and effort but is really worth it in the long run.

It is too hard to get our heads around all of the Work-As-Written stuff (Procedures, Work Instructions, etc.):
Yes, and a lot of this is because not everybody understands that Work-As-Written does not always match Work-As-Done and that Work-As-Written is often created without the end user in mind. A consolidated, in-depth study of the 'accuracy' of the Work-As-Written stuff can really help here. To do this, simply get a small team together, including some leaders, and grab all of the rules, procedures, work instructions, directions, guidelines, and standards that apply to a task. Lay them out on a large table and go through them in fine detail. See if it is possible to actually follow them. In the vast majority of cases, you will see that it is not possible to follow all of the rules all of the time. Then, of course, you need to fix them.

I know how work should be done and I know that people do not always follow the procedures. This is okay, as long as they do not hurt themselves:
Sorry, but this is the wrong attitude. You need to make sure the procedures are accurate. And are able to be followed. They need to be easy to follow correctly and difficult to follow incorrectly. And just because someone gets hurt or does not get hurt must not change the way you should view compliance.

As a Manager, I do not always follow my own rules:
It is great that a leader recognises this. The secret now is to translate this recognition into open conversations about following rules with those others in the business that the leader may expect to follow the procedures. It is also about having a willingness to start to eat the elephant by creating accurate and easy-to-use-correctly procedures, work instruction, and other system requirements.

It is too hard a job to rationalise all of our Work-As-Written procedures and systems. Some of our procedures are complex and some are simply complicated:
Start by fixing the procedures that cover the critical tasks and work out from there.

Understands Their Own and Others' Expectations

We all have expectations of ourselves and of others. We do not, however, always let them get too close to the surface of our thoughts, words, and actions.

The boss's expectations of us cannot be achieved:
The simple and yet daunting remedy for this barrier is to implement a 'Speak Up' programme. The intent of such a programme is to encourage worker-level employees to talk about issues in the workplace. If this doesn't work quite the way you intend it to, which is often the case, convert the 'Speak Up' programme to a 'Listen Up' programme. This is aimed at encouraging leader-level employees to listen to what others say. It can easily link into the Listen Generously work.

The business expects things like procedural compliance, which we cannot always do because our procedures are not all 'accurate':
This is more about the leaders' views on the accuracy and usability of procedures rather than expectation setting. Tied in with a 'Speak Up' or a 'Listen Up' pro-gramme, try to get the leadership team to talk about the accuracy of procedures rather than just following them, especially when following them is often not possible.

Our bosses tell us what to do and we do it, so we know what they want. We know their expectations because they tell us what to do. They don't need to understand our expectations of them:
Have a thought about what you expect from your boss in addition to what expectations you have about yourself. Ask yourself how you should or would like to behave in these circumstances. Talk with your boss about what you feel. If you do it in terms of your expectations of yourself first and then about the expectations you have of them, you may be surprised at the level of conversation that can ensue.

We all have to sign a 'personal pledge' saying we will always follow all the proce-dures, but we all know that we can't. The bosses know this just as well as we do:
This is absolutely about authentic leadership. In the past, I have simply refused to sign off on procedures that cannot be followed. When you do this, you need to do it in a diplomatic way, explaining why you cannot sign the personal pledge. Talk with your boss about what you feel. As in the above barrier, if you do it in terms of your expectations of yourself first and then about the expectations you have of them, you may well get a good hearing.

As bosses, we can be guilty of saying one thing and doing another – we do not always follow procedures either. One example that springs to mind is our Management of Change processes. We rarely follow those:

Get the leadership team to have a dedicated meeting to discuss authenticity in their leadership, particularly with respect to messages and behaviours. Saying one thing and doing another is clearly visible to your teams and can destroy credibility very quickly. The best remedy is simply to stop it. Either follow your own systems and procedures or change them – you have the power.

We keep getting told that safety is our responsibility – is that the boss's expectation? I am not sure:
Ask them. And then have a conversation about what they think safety actually is and where the various bits of responsibility lie. Pick your time for this. A performance review appraisal is a good time for an authentic conversation on what each party's expectations are of the other.

The boss's expectations are just words. They don't really mean it, but they can say they have told us:
If you are a leader, you can change this by 'adopting an authentic leadership approach when leading others'. You and your behaviours are very transparent to those you lead, even if you think they are not. You can behave like a great leader until you become a great leader as long as you do it from the heart – authentically.

I do not have any expectations of my boss, and he has never asked what they are anyway:
Take some time to think about your 'Why', what drives you, what your ethics are, and of course what your expectations are about how you want to, should do, and will behave. Then think about what expectations you have of your boss, and then, at the right time and place, talk about it with them.

Understands the Limitations and Use of Situational Awareness

Situational Awareness is one of those fickle concepts that gets a lot of airplay. It is a stock 'root cause' for incidents. Loss of situational awareness and poor CRM (Crew Resource Management) seems to be a common go-to cause for aviation incidents also. To me, this is about as useful as citing 'lack of hazard awareness' as the cause of a workplace incident. It is clearly not a cause of anything. It is often obvious after an incident – in hindsight – that someone has lost situational awareness or was not aware of the hazard that has just manifested itself. It does not help us learn much. And learning is what incident investigations are all about as I have talked about earlier.

When someone misses something in the workplace, they have lost situational awareness. This is a fact and it is obvious. Why are there limitations to Situational Awareness?
Humans are simply not able to see and understand everything that is going on in their part of the world. We constantly prioritise what we see, sort out which bits are important, and keep an eye on them. This is often at a subconscious level. We see things but do not always register what we see. Watch 'The Invisible Gorilla' and 'The 'Door' Study' on YouTube to get more on this. Both clips are by Simons and Levin.

Workers must keep an eye on everything around them. That is their job:
Again, this is not possible. The remedy here is to help the workers understand which bits of the task or their surroundings are important and that they need to keep an eye on. Acknowledge to them that it is highly likely that they will not see things that they do not expect to see. This is also a reason why we need to focus on trigger steps and critical steps in tasks as these are steps that they absolutely do need to keep an eye on.

I can't keep an eye on everything, but my boss tells me I need to:
You are absolutely correct here: you cannot keep an eye on everything. Before starting the task, stop and chat with your workmates. Talk about the trigger steps and the critical steps and what you need to keep an eye on. Then hold each other accountable to keep an eye on those things. If you are a leader who is telling their workers to keep an eye on everything, you need to stop this and work with your people so that you all know which bits of the task are critical in terms of situational awareness and which bits they need to focus on. You need to build a shared mental model of what is important and what is less so.

When I am super-focused on a task, I do not see what is going on around me:
This is a classic 'Invisible Gorilla' scenario. The more intensely you focus on something, the less likely you are to see other things. Build in stops and pauses as you undertake the work so that you can take in what is going on around you.

Do you want me to keep an eye on the weak signals or the steam train coming down the track at me?
Both actually. But it is a good idea to keep a keen eye out for a train if you are working on the tracks. Build resilience into the task that you are doing so that you have already thought about what could go wrong. This way you will be more likely to see it coming. Additionally, especially when you are good at a

task, you can shift your (limited) focus capability to the weak signals that could manifest and catch you out or warn you of an impending change.

Nobody has ever asked me about Situational Awareness. Can I use a different language?
Absolutely. This should be the approach with any of the elements in *Essentials of Safety*. It is the concepts that are important. You could talk about 'keeping an eye on what is going on around us', or 'focusing on those few things that could make a difference here'. Whatever language works for you is fine, as long as it conveys the intent of Situational Awareness, its use and limitations.

I do not know how to ask people about Situational Awareness or mental models. How can I use language that they understand?
As per the above remedy, use language that works for you and the people you are talking with.

When is the best time to ask people about Situational Awareness or mental models?
At any time really. But especially when you are out and about in the field talking with workers who are at risk and are doing the work. A team and a leader having a shared mental model of the work being undertaken – what is going on around the workers, what to keep an eye on, and what we need to get right – are all excellent and easy conversations to have.

Overall, encourage teams to stop and think about what they need to keep an eye on *before* they start a task and at periodic points during the task. This is the way to keep the idea of situational awareness alive and the team focussing on what they need to pay attention to.

Listens Generously

Listening is the most important skill a leader can have. This maxim does not seem to be well known. This is a pity as it leads to what I believe is the biggest barrier to effective and generous listening. If you ask a group of people in a workshop to write down where they believe they sit amongst their peers on a scale from 'poor', 'below average', 'average', 'above average', 'very good', and 'excellent' in terms of their listening skills, you will nearly always get a skew towards the upper end of the scale. I have tried this numerous times and tend to get the majority of people balanced somewhere between above average and very good, with a few outliers covering the rest. This supports the idea that the majority of people think they are already good listeners. When leaders get promoted, the behaviours and practices that the leader exhibited at the lower level can be perceived as the right way to behave as a leader in their new role. This includes how well they listen to others.

This could be viewed as those leaders having a fixed mindset around their listening skillset. I prefer to give them the benefit of doubt and explain the behaviour as a lack in the leaders' leaders. By this I mean that if their leaders had a good understanding of what great leadership looked like and developed their people in that light, they would have had a better chance of ensuring that generous listening played a part in the development plans as their people moved up the leadership ladder.

This points to the fact that the skill of generous listening as a part of a leader's development plan should not be restricted to those newly entering the ranks of leadership, but should also encapsulate anybody who has any leadership require-ments in their role. To me, this is just about everybody in the business.

Let's look at some specific barriers and remedies and then we can look at what I feel is the simple remedy for all the barriers.

I haven't got time to waste listening to people who do not know what they are talk-ing about:
Think about the fact that, despite your brilliance, you do not know everything. You never know when a person who you may think does not know what they are talking about comes up with a thought-provoking, if not brilliant left-field, idea. It is far from a waste of time listening to people offering their thoughts and ideas. Even if they are not experts in the field they are talking about, the fact that you are listening sets up a culture of trust where people know they can come and talk with you and that you will listen. This is a good leadership trait to develop and practice.

I have an opinion and if I do not interrupt others and speak up, my ideas will not be heard:
If this is happening in a meeting, install a listening coach. Their job is to give every meeting participant feedback at the end of the meeting as to what the quality of their listening was like. Also, you can listen to what others are saying and when the topic is similar to your views, you can say something like 'I agree with John on this, and I feel that …'. The 'I agree with John' bit may attract the attention of the talker and they may well then give you space to talk.

In leadership teams, the example set is that it is everyone for themselves – interrup-tion and talking over people is the rule rather than the exception:
In addition to establishing the function of a listening coach within meetings, establish no-interrupt rules, and that people can only add to what the speaker is

talking about, not to just throw in their unrelated thoughts. You may also have to establish a 'time keeper' role to encourage people not to hold the floor for the whole meeting.

I know what the answer is so I do not really listen:
This is a dangerous habit to fall into. It usually comes across as disrespectful, even if that is not the intent. The remedy is to stop all other tasks. Close the lid of your laptop. Turn your phone over, and pay all of your attention to the speaker. You never know what brilliance may manifest from their words.

I need to get emails done and so I often listen with one ear to people in my office while I read my emails:
Despite popular opinion, humans are very bad at multitasking, especially if the tasks require cognitive work. If your emails are important and the people talking with you feel their words are important, you simply cannot give sufficient mental power to both tasks at the same time and be effective in either. The best remedy for this one is to stop it. You need to make a decision as to which is the most important activity at the time. And then stop doing the other one. This may entail you suggesting another time for the conversation or a delay in the reading of the email. It is your call, but you can only choose one.

Nobody has ever told me that I do or do not listen properly:
This is where the role of a listening coach is valuable in meetings. Also consider including listening generously into people's leadership development plans and routine performance appraisal processes.

I think that my listening is above average, although my partner at home may not agree with that:
By definition, only half the population can be above average. Listen to your partner. Ask to have listening generously included in your leadership development plans and routine performance appraisal processes.

I always like to express my opinion as a leader first and then see if people agree or not afterwards, that way, we get alignment nice and quickly:
You may get alignment, but you risk losing a great opportunity to see what others think. A better approach on the journey to excellence is to ask them first before you give your views. In order to avoid a group think, it is even better to get them to write ideas down first and then to talk about them.

You cannot learn to be a good listener, you either are or you are not:
This is a fixed mindset view. You need to move this to a growth mindset. Check out the barriers to mindset earlier in this chapter.

So, in summary, the remedies to consider are as follows:

■ In meetings, implement a listening coach.
■ Include listening generously into people's leadership development plans and routine performance appraisal processes.
■ Practice building what you want to say into what the last person speaking was talking about – build on, rather than break in.

Plans Work Using Risk Intelligence

We need to be both efficient and thorough all the time, especially when planning:
I use the example of NASA and the investigation that followed the 2003 Columbia shuttle incident when explaining the concept of the Efficiency-Thoroughness Trade-Off during workplace incident investigation training. When you are planning an incident investigation, you need to think about how many people will be involved in the team, how many people you need to talk to, and amongst other things, how much time and effort you will exert in the process. You have a choice. You could choose to be thorough, just like NASA after a space shuttle disaster: the Columbia shuttle incident investigation cost NASA about $400 million. In addition, it tied up thousands of people for a year and wrote reports that totalled over 650 pages, including the crew survivability report. The other approach is for you to be efficient. To do this, get two people together in a room for an hour or two and write an incident investigation report that sounds about right. In the real world, it is always a balance. I have applied the ETTO principle in writing this remedy. I could have gone on for pages and pages, but I needed to strike a balance that gives you enough information to get it, without overloading or boring you. I hope I got it right.

We do not teach our people how to identify hazards, yet alone about Risk Intelligence:
I believe that you should absolutely do both. Combining them together is an effective and efficient way of helping people learn. The training sessions do not have to be long but need to be simple and very practical. There are some great Virtual Reality-style hazard identification training programmes available. Mix them up with some education on how Risk Intelligence fits into their world and how they can get better at it.

I do not want my team to be looking for negatives (a suitable wariness for the effectiveness of controls):
I have said many times that the focus needs to be on making sure things go right. Having said that, I think that having and nurturing a suitable wariness for the effectiveness of controls is a bit different from looking for negatives. What you need to promote is for people to stop and think, to challenge whether in the situation and workspace they are currently in, the controls they plan to introduce will work as they are expected to, or whether something may not quite work as planned.

We want to encourage a strong focus on 'faster, better, cheaper, safer'. This seems at odds with what you are talking about?
In a similar way to efficiency and thoroughness in the ETTO principle, you cannot easily serve the masters of 'faster, better, cheaper, safer' at the same time. It is always a trade-off. NASA learnt this the hard way after the Columbia accident. As long as you are clear about the fact that it is always a compromise and you keep going back to what you are going to achieve and then nurture a suitable wariness for the effectiveness of controls, you will be fine.

Planning is a function that planners fulfil. We expect the people who do the work to work the plan:
This is okay, as far as it goes, at least for the large planning activities that planners do. There is, however, an endless amount of planning that goes on at the individual level that must be understood and thought about. Working the plan requires planning by the individuals working the plan. What we are trying to say is that at an individual level we need to maintain the Efficiency-Thoroughness Trade-Off, have a good level of Risk Intelligence, and maintain a suitable wariness for the effectiveness of controls. This applies just as much to task-based planning as it does to the planning the planners do for production and maintenance activities.

Our people are busy enough as it is without adding 'time to think' to their schedules:
I am sorry, but this is woolly thinking. Leaders at all levels must make time to think. At a strategic level, thinking time is vital and time should be allocated in your daily schedule for thinking. At a more transactional level of leadership such as a frontline leader or a worker, thinking time is just as important. We need to think about why we are doing what we are doing. We need to think about what needs to go right, what might go wrong, how we view the work, what our mental model of the work is, what elements of the workplace and the work do we need to keep an eye on. This all requires thinking time.

If we remain preoccupied with failure, we will never get any work done:
There are a number of ways of thinking about being 'preoccupied with failure'. Other related terms include 'chronic unease' and nurturing a suitable 'wariness for the effectiveness of controls'. The idea is not to go overboard with any of them. Do not be so preoccupied with failure that nothing gets done. Do not be so chronically uneasy that you can no longer function. Do not be so wary of the effectiveness of controls that you trust none of them and all work stops. Once again, it is about maintaining a balance, not forgetting that things do go wrong, controls do fail, and we want to think about this a bit as we carry out our work.

You cannot wrap our people up in cotton wool:
Absolutely not. I agree. Again, it is about maintaining the balance. We must remember, however, that there is one guarantee in a workplace that contains humans. You can guarantee that humans will fail. Our job as workers and as managers and leaders is to allow people to fail safely. This is why we need drop zones around people working at height. Humans drop things.

Our industry has got hazards in it. Worrying about them does not help:
There is a difference between worrying and having a suitable wariness. A level of unease is very healthy. Being constantly vigilant is hard work. Spend time before starting a task to talk about what must go right and what might go wrong.

If we spend time focussing on putting extra controls in place, no work will get done:
We need to have sufficient controls in place to allow us to fail safely. Some controls can be about doing things in a way that ensures it goes right, rather than creating barriers to failure. Approaching controls as 'preventive' controls as well as 'corrective' controls can help maintain the balance between overcontrolling work and allowing risks to be too high. Once again, it is all about maintaining the balance.

We consider ourselves a High Reliability Organisation (HRO) so we already do this:
If you consider yourself a High Reliability Organisation, you probably are not one. There is a lot said about this. If you actually mean that you are constantly striving to attain HRO status but recognise you have a way to go, then everything in this book still applies. If you truly believe you are already there, then you need to spend some time contemplating how complacency might be impacting your focus, leadership views, and behaviours.

We should celebrate success – the number of days since our last injury is a great thing to measure – it tells us how safe we have been and keeps focus for our people at all levels:
There is lots of evidence to suggest that driving the numbers does absolutely nothing to drive sustainable safety. The number of days since your last injury is of no use in predicting when the next one will occur and it tells me absolutely nothing about how safe your workplace is. We need to explore and understand how the work is being done on a day-to-day basis to assess whether we have a safe workplace or not. Measuring the outcome of work is not a measure of safety. In my learning study training workshops, I show a picture of a young man shaving with a hunting knife. He has been doing it for ten years without any serious injuries. Over 3,650 days without a recordable injury. An injury rate of zero. Does this give you reassurance that it is being done safely? Absolutely not.

The idea of 'Is preoccupied with failure' seems at odds with safety being all about the things going right?
It does seem that way upon a first read. However, like most of what I talk about in this work, it is a question of focus and of course, balance. Initially we need to do what we can to set up the work for success. We need to understand what we need to do to ensure things go right, and only then, as the work is being done, do we need to keep an eye on things in case something goes wrong.

Controls Risk

As long as we do not have workplace incidents and injuries, this means that we must be controlling the risks:
Past performance is no guarantee of future performance. We need to understand how we have managed to not get any injuries before we can claim that the lack of injuries means that you are doing safety well. I suggest doing a comprehensive study to explore what the drivers of the lack of injuries are. You need to deeply understand Work-As-Normal, any differences between Work-As-Normal and Work-As-Written, along with what underlying organisational factors may be driving near miss and close call incidents. You need to understand whether any incentivisation of safety outcomes may be influencing reporting. You can only tell whether you are controlling risks by being out there with the workers exploring what is going on in their world.

We have an injury rate of less than 3 per million man-hours. We are controlling the risks in the workplace – this is a given by our results:
Rare events happen rarely. I use some highly spurious statistics to tell a story: having a Total Recordable Injury Frequency of 3 means that three times out of

a million we do not end up with an injury. This means that 999,997 times out of a million we do not have that level of injury. This means that 99.9997% of the time we get it right. A Total Recordable Injury Frequency of 20 would mean that you are getting it right 99.998% of the time. There is not a lot of difference there. Rather than saying that we are controlling risks in the workplace because we have a Total Recordable Injury Frequency of 3, it is better to try to understand the underlying factors which result in us getting it right 99.9997% of the time. You can guarantee that your people are not following your procedures 99.9997% of the time, so something else is going on. What is it? Understand this and you can then talk about what you are doing right on a day-to-day basis that demonstrates that you are running a safe business.

We have procedures that are critical. In ones like 'Isolation' everything in the procedure is critical:
The goal of a critical control or critical step in a procedure is to make sure that the users of the procedure know that that particular task requirement *must* be done in a certain way and that it is more important than other requirements of the task. I am pretty sure that there are requirements in your isolation procedure that, whilst important, are not as critical as others. An example could be the process for cutting off someone's personal lock if they have gone home or how often the procedure needs to be reviewed. There should only be a few requirements of a procedure that are critical. I like to think of building procedures that contain two types of requirements; 'framework' requirements and 'critical' requirements. This is talked about in Chapter 1. As a reminder, 'framework' requirements align to the concept of 'freedom within a framework'. This is where work teams and/or individual workers decide how the work will be done within a framework. The framework is set with clear guidelines and context but limited information on exactly how to achieve the task. The 'critical' requirements of procedures are critical in the sense that they are a step within a task that must be completed in a very specific way, accurately and repeatedly each and every time the task is done. The idea behind a critical requirement is that failure to follow the critical step in the way prescribed could result in a fatality or other similar significant incident.

We have never thought of 'trigger' steps. Adding this on top of 'critical' steps or controls will add complexity and confusion:
I believe the addition of trigger steps helps the worker be more aware of what they are doing and reminds them that the trigger step cannot be undone. It also

aids in the situational awareness aspect of the work. It ties in very well with tele-graphing deliberate action, which is where workers verbally describe the tasks they are about to do.

How can our people focus on all of this stuff, as well as not getting hurt as they control risks? For example, trigger steps, critical steps, ETTO, line of fire, SIMOPS, workplace anomalies, telegraphing deliberate action, safety 'routines', shared space, having a wari-ness of the effectiveness of controls, the context of the task – the 'Why?', making 'time to think', changes to the task or surroundings, situational awareness, preserving options, applying procedures thoughtfully, even getting their heads around the hundreds of pro-cedures and systems that exist in our workplace. This is an impossible task:

When you put it all down as a list, it does look like a lot. What you have actually done is to list just a few of the things we all need to do all of the time at work. If you tried to list all of the things that we need to do just to get out of bed in the morning, have a shower, clean our teeth, get dressed, have breakfast, and get to work, it would be the size of a book. The secret is to consider all of these things as just things that we do. They need to be developed into routines that are as familiar to us as cleaning our teeth. If you watch someone working, you will find that they are already doing many of these things most of the time. We are just calling them out as essential. For example, workers always apply the Efficiency-Thoroughness Trade-Off in all that they do. They watch out for things in the line of fire. They keep an eye out for what others are doing around them (Situational Awareness and SIMOPS). They watch their critical controls. The number of new things in the list and in this book is actu-ally quite small. It is about ensuring our people, especially our leaders, are aware of them and keep them front of mind and in their routines on a daily basis.

Applies a Non-Directive Coaching Style to Interactions

I do coaching all the time – I am always telling people what to do:
There is a difference between telling people what to do, mentoring, and coach-ing. Coaching is more about asking questions, listening, and responding to people, all for the purpose of helping the player be the best they can be. Coaches do not have the answers, the players do, so telling players what to do is only a very small component of coaching.

I have never been trained in coaching:
If you cannot get your boss to get some coaching training for you, a great place to start is the following list of resources from the bibliography at the end of this book:

John Whitmore (9); L. David Marquet (114); Myles Downey (24); Max Landsberg (54); Ronald Heifetz (105); Sharon Parks (46); Carol Wilson (66); David Rock (70); Kerry Patterson, Joseph Grenny, Ron McMillan and Al Swizter (16); Carol Dweck (56); Ian Long (88).

Coaching equals Mentoring equals Leading:
Sorry. Coaching does not equal Mentoring does not equal Leading. The principal difference between coaching and mentoring is that with mentoring many of the answers lie with the mentor. Whilst with coaching, nearly all of the answers lie with the player. Leadership is all about authentically caring and influencing others through relationships. Leadership is not a hierarchical thing and leadership is not about controlling direct reports. We need to always remember that leadership is something we do with our people, not to our people. This is especially true for intent-based leadership.

I do not understand what you mean by 'non-directive' coaching:
Coaching can be either directive or non-directive. Directive is more about giving advice, instructing, telling, and pushing solutions to the player's problems. Non-directive coaching is about asking questions to raise awareness, listening to understand, and helping players solve their own problem. The most effective coaching to help people be the best they can be is about an 80:20 ratio between non-directive and directive coaching styles.

It is much easier as a leader to tell people what to do – I have not got time for coaching – It takes too long. I know what needs to be done:
Give a man a fish, or teach a man to fish. This saying sums up the dilemma you talk about here. In the immediacy of the moment, it seems a lot easier to simply give an instruction. If you do this, you will forever be spending time giving instructions. If you apply a coaching style, the worker will soon learn what needs to be done, will have the context and purpose of the task, and will not need to be instructed each and every time they come across the same or a similar problem. Coaching is an investment in freeing up your time.

There are many coaching models out there and they don't all match the idea of non-directive coaching. I want to use another one:
I was taught the GROW model about 12 years ago and have been practising it ever since. This was after looking at many coaching models. I have found that it is easy to apply, seems very suited to workplace coaching and can easily be adapted to a style within managing and leading in addition to use in a pure

coaching activity. Any coaching is better than no coaching, so if you are used to another model, use it. As long as your intent is to help the player be the best they can be, you will not go wrong.

I do not see the benefit to our workers if we coach, rather than guiding and telling them what to do, including explaining the 'Why?' of the task:
As the answers to people's problems come from themselves when they are coached, the learning is sustainable. This is not always the case when we simply tell them what to do. Explaining the 'Why' of course helps but it is better to use a coaching style or coaching itself in your endeavours to help the players be the best they can be.

The benefit to our leaders is not clear if they coach or use a coaching style in their leading and managing approaches:
This is the million-dollar 'Why' question. It is often expressed as 'What's in it for me?' The answer lies in the strength of your team, in the increased ability of them being able to solve their own problems, in their increased understanding of the broader context of activities, in their own development and learning, and, of course, in the freeing up of your time. It is a lot easier to lead a team that is full of competent and excellent people than a bunch of people who ask you questions 24/7.

Has a Resilient Performance Approach to Systems Development

Systems development is not my responsibility:
By systems development we do not just mean the high-level organisational-level systems. We also mean the lower-level procedures, work instructions, and Task-Based Risk Assessments that all leaders get involved with. Many middle- and lower-level managers have a large role to play in the development of these system tools. If you are a worker, you also have a role to play in system development. You are the one who will need to use those systems and so you have a responsibility for making sure the systems you use are accurate and are able to be followed. You then have a responsibility to talk to your manager if those systems cannot be followed.

Document Control, safety people and engineers create the systems. Our leaders just make sure people use them:
Your organisation has a huge problem if you are leaving the creation of systems up to safety people, engineers, and document controllers. We need to allow expertise to lie where expertise lies. In the vast majority of cases that expertise

lies with the people who do the work. They know what works and what does not work in the real world. They are the ones who need to use the systems. They need the deciding vote on whether a specific procedure or system is any good or not. The remedy here is simply to change the system so that the right people are involved in the creation and approval of systems.

I cannot influence what goes into the site and corporate systems:
If you are the user of a system or have an accountability to ensure its use, then you will also have some level of influence over those systems. If a site or corporate system is not workable, you have an accountability to speak up about it. Gather like-minded leaders and make some noise. I have seen some amazing results with some persistence and determination.

Our safety systems are already in place. Why fix something that is not broken?
Every business I have worked with has had some sort of safety system in place. Every business I have worked with has had a safety system that needs work and is not anywhere near perfect. The vast majority of safety systems that I have seen are so complex they cannot easily be understood or followed. They routinely contain double binds for workers or are written in engineer-speak that workers do not understand. I recommend undertaking an analysis of the accuracy of the safety systems. By 'accurate' we mean – as discussed in Chapter 1 – 'Accurate includes being effective and representative of the organisation's collective wisdom (including the end users) on the best way to accomplish the task'. You will most likely find that you need to do a fair bit of work in this space.

Fixing our systems is an adaptive problem – a problem without an easy answer:
This is a common barrier to the creation of great and effective systems. Solving adaptive problems requires an approach that challenges the status quo and the way you normally approach problem-solving, and in this case, how you normally create systems and system elements. You need to absolutely re-think what you are trying to do, what the purpose of the system is, how it interrelates with other systems, what the system must, or could, look like, and why. You also need to think about who must be in the room as the systems are developed (the end users for example). Most of our workplaces are complex socio-technical environments and so the systems that drive them need to take that into consideration.

Our system is bigger than Ben Hur. This is all in the too hard basket:
Q: How do you eat an elephant? A: One bite at a time. Start with the critical activities and the systems that drive them and go from there – with persistence

and urgency. One way to ensure you do not lose momentum is to make the work a topic of conversation at the highest level in the organisation. It needs to be of continuous interest by the senior leadership team of the business.

On the Safety Oscillation Model: Our safety people and our leaders should know what to emphasise or minimise to drive safety outcomes:
The idea of the Safety Oscillation Model is to make the connections and influences more visible. Sometimes leaders, whether safety or operational, get tied up in the detail and miss the wood for the trees. The Safety Oscillation Model is intended to help them take a step back and think about what is driving what. It can be viewed as an alternative to the classic safety value driver tree. The content is nearly always known but it helps to be reminded of the details now and again.

Adopts an Authentic Leadership Approach When Leading Others

I keep getting promoted so I must be a good leader:
I know some very senior managers who have not got a leadership bone in their bodies. There are other people I know who are brilliant leaders and have not got any promotions at all. In my view, the concepts of leadership and hierarchical level in a business are neither correlated nor causally related. When you are considering an employee for promotion, just as when you hire someone, an element of the decision-making process should include an assessment of their leadership skills and leadership potential.

Our leaders and managers are chosen partly because of their track record and experience, not their leadership prowess:
I think we need to be careful not to be confused between leaders and managers. Managers are chosen by more senior managers. Leaders are chosen by their followers and those they influence.

My leader thinks he is a good leader and tells us so quite often. We do not think so:
Again, this barrier seems to have come from the common confusion between leaders and managers. I read this barrier as '*My manager thinks he is a good leader and tells us, but we do not think so*'. Self-declaration of leadership is not any kind of a measure of the quality of leadership. Leadership is felt by followers and those the leaders influence. This is an issue that the leader only can address. If you feel, as a manager, that you are a good leader, do not trust that feeling. Explore it with those you lead. Seek out feedback on your leadership and see if it agrees with your perceptions of yourself. And work from there.

Our leader says the right things, but does not do them himself:
This is a classic barrier to great leadership. It sounds like the leader is not a truly authentic leader. This is when you need to talk with your two-up manager about the impact your manager is having. If you are the leader who is doing this, then you need to take a look at what you are doing in your behaviour and ask whether it aligns with your 'Why'. Ask whether the way others see your behaviour is as you intend it to be or whether there is a misalignment of some sort. And work it from there. How others see you and your behaviour is what makes or breaks you as a leader, not how you feel about yourself.

I behave like a great leader and so people believe I am one. What more do you want?
The trick here is to move from just behaving like a great leader to authentically becoming a great leader. There is nothing wrong with behaving like a great leader until you become a great leader but it needs to be done with the purpose of becoming a great leader for the benefit of others and it must be done authentically. Great leadership is not about you and it is not something you do to others. Great leadership is something you do with other people and if your behaviour is not authentic, it will get you unstuck eventually. Reflecting on your 'Why?' will also help greatly here.

I need to retain overall decision-making authority. This is my job as a leader:
The simplest remedy for this problem is to consider the words of David L Marquet in *Turn the Ship Around*. It is a fabulous story about Marquet, who felt this way as a nuclear submarine commander. He knew it was wrong and so changed his leadership style and ultimately the way leadership is done in the USA navy. *Turn the Ship Around* is a great read and a meaningful story about intent-based leadership. Intent-based leadership is all about giving decision-making control to those who have the information. It is about giving the decision-making authority to those with the expertise. It is about people setting the context and intent of the work and allowing those doing the work to tell others how they intend to do it. It is an extremely powerful way of building trust and great leadership.

I am the leader. I have the expertise, or I would not have been given the role I have:
That is not always accurate. You may be brilliant from a technical perspective. That does not necessarily mean you are brilliant at leadership *per se*. Think about the statement 'I am the leader' as compared to 'I am the manager'. Being the leader is not a self-declared position. It is a title given to you by those you lead.

Spend some time thinking about why being a leader is more powerful than simply being a manager.

I understand what is expected of me as a leader and I will toe the line, but it is not exactly how I would normally operate outside the job. I have a 'leader' hat on and a 'real person' hat on as needed:
Being authentic in your leadership will result in your leadership behaviour outside of work being exactly the same as your leadership behaviour inside of work. I suggest exploring and practising authenticity in leadership in all aspects of your life.

Failure is failure is failure, and it is important that I am not seen to fail as a leader:
Great leaders open up and admit failure and view it as learning and not something to be precious about. Authentic leaders will often talk about how past failures have helped them develop their leadership style. Talking about who you are, what has worked in the past, and what has not worked in their leadership journey is a great way of building authentic leadership.

Leadership is something that we do to people. By definition, we 'lead' them:
This barrier is quite common and it shows a lack of understanding of what leadership actually is. Leadership is something you do *with* people. Leadership is not something you do *to* your people. A leader is an integral part of the team. You need to learn that leading is not managing is not mentoring is not coaching.

In summary, I am sure I have missed out on a lot of other barriers to the implementation of the essential elements of safety in this chapter. I will have also missed out on a tonne of possibilities for remedies to those barriers. My intent was to give sufficient information for you to consider and build on as you explore what works for you, what your own personal barriers are, and your own ways of overcoming those barriers.

Chapter 4

The Essentials of Safety
as a Driver of Learning

You see, but you do not observe.

A Scandal in Bohemia, **Sir Arthur Conan Doyle**

I have always pushed the idea that the conversations we have after a workplace incident should be the same as the conversations we have before an incident. This is what prompted me to explore whether it was possible to use the fundamental ideas of – and the specific elements of – the *Essentials of Safety* as lenses for exploring learning within an investigation method after a workplace safety incident in addition to using them for Created Safety before something goes wrong.

Whatever investigation method you use in your business, I believe it would be beneficial for you to use the *Essentials of Safety* elements as an adjunct to that process. Thinking about how each element played a part, or could have played a part, in the incident will give you a different perspective as you try to unravel the incident and its drivers. The same is true when you are exploring differences between Work-As-Done and Work-As-Written on a day-to-day basis. Using the elements as lenses to observe work through and as prompts for learning will help.

As we have said throughout this book, the essentials of safety are a set of individual characteristics, distinctions, attributes, or traits that permeate through the workforce at all levels. The essentials talk to each viewpoint of the *Individual*, leaders and *Leadership*, the *Systems* we use, and the *Cultures* of the workplace.

DOI: 10.1201/9781003181620-2

If we accept that getting these elements right and embedded in 'the way we do things around here' will help drive towards a state of Created Safety, then the logic says that we can use these elements as a basis for creating conversations to explore what we can learn from either work as it is done on a day-to-day basis or when we have an incident.

I asked myself whether it is valid to consider if we can use the *Essentials of Safety* elements as a basis for learning studies after something goes wrong. I am confident that the answer is yes. If the presence of the elements exists within the 'thinking' and 'doing' across the various workgroups and individuals within the business and this leads to, supports, and otherwise helps create safe work, then the absence or partial embedment of those attributes or elements could lead to a state of 'not safe work'. I am not going for the strict Newtonian cause and effect level of causation here. Rather I am talking about a level of contribution that, when combined with and interacting with the other elements, conspires to result in an upset or workplace incident as an outcome of the work.

In the main body of this chapter, I will offer thoughts and idea prompts for each of the elements that are worthy of consideration within your workplace incident investigation process, whatever that may be. Towards the end of this chapter, I will offer an example exploring an investigation method based on my previous experience and work, including the main ideas from *Simplicity in Safety Investigations*. I have expanded this to include the ideas and elements of *Essentials of Safety*.

Before we explore the possible contributions of each of the *Essentials of Safety* elements, I feel that we need to dig into what learning is all about as compared to sharing findings or simply understanding an incident. Businesses seem to be much better at sharing information after workplace incidents than actually learning from them. The difference between sharing and learning is enormous. Sharing could be simply creating a safety alert and sending it out to all points of the compass and then asking leaders to go through it with their work groups and teams. This may result in an increased level of awareness of the incident and its drivers in the short term but rarely yields changes in fundamental beliefs, behaviours, and practices in the workplace – learning.

Learning is all about the relatively permanent behavioural, organisational, engineering, and cultural modifications which take place as a result of experience. This change in behaviour needs to be long-lasting for learning to be considered as having been achieved and must come as a result of experience. This implies that telling people about something they need to learn is not a very effective method of enabling learning; neither is sharing PowerPoint presentations or reading an incident report to your team during a pre-shift meeting. We need to create a learning process that brings an experience to the fore. Stories can work and are most powerful when they are given by those directly involved in the incident that initiated the need for the learning. We need to do work in the embedding of learning such that we are updating people's mental models of the way they work and what they believe in relation to their work.

What follows is a list of prompts that may trigger ideas and thoughts that help you understand an incident or day-to-day work. It should be read in conjunction with your incident investigation model. For example, when considering what data or information you may want to see as a part of the data-gathering step of the investigation.

Understands Their 'Why'

■ The level of clarity regarding whether those directly involved in the incident knew why they were doing the task:
  – Whether those involved know where the task fits into the broader picture of their department and business.
  – Whether they know where the task fits in with the business context and strategy.
■ If 'drift' is present in the way the activity was done on the day of the incident, try to understand the level of clarity as to why the task was required or be done in a certain or specific way:
  – Here we are exploring if the observed drift was driven by a lack of understanding by those doing the work with respect to the specific requirements of the task. For example, in the case of the *Herald of Free Enterprise* ferry disaster, we see that over time the drift-related increase in production pressure to leave Zeebrugge early meant that 'harbour stations' were routinely called before the bow doors were checked as shut by the deck officer. The crew did not realise the vital nature of ensuring the bow doors were physically shut prior to sailing.
■ The presence and level of 'Why' discussions during training and/or employee performance reviews and appraisals:
  – By looking at a number of training packages and talking to various people, we can determine whether the context of the thing being trained in is explored during the training session as a normal practice. This can be checked by looking at the records of measurements

collected as a part of assessing and measuring activities discussed in Chapter 5.

■ The level of discussion of the 'Why' of a job during training of the specific process or procedure associated with the incident:
  – This is more specific than the previous one and relates just to the task being undertaken at the time of the incident.
■ The level of interest of people understanding their 'Why':
  – Do people really know why they do what they do? Is it obvious that this element has been discussed and explored with people in the business or not?
■ The amount of time leadership spends exploring the 'Why' of roles and tasks with workers:
  – This can be estimated during conversations not with the leaders themselves, but by talking with the workers doing the actual work.
■ Process and practices around ensuring teams understand task assignment and intent:
  – Conversations need to be held with those assigning tasks and also with those receiving the task assignments in order to understand this aspect.

Chooses and Displays Their Attitude

■ How people view their tasks, their roles, the business, and their leaders:
  – Love, hate, boring, vibrant, fun, just for the money?
■ The levels of cognitive dissonance within the team based on how they view what they were doing at the time:
  – What are the competing feelings in people? For example, working as a butcher whilst being a vegan.
■ Level of understanding of 'attitude' as a concept and as a choice:
  – What conversations occur during induction processes, leadership conversations, routines, training activities, annual performance reviews, and appraisals to help people understand the impact of how they approach and view their work?
■ The balance of attitudes across the work team/s:
  – We need to talk with a number of others to get to this. This is okay as we need to talk to a number of other work groups in order to understand Work-As-Normal anyway.

Adopts a Growth Mindset - including a Learning Mindset

- Where the local work team and the broader leadership sat on the continuum from fixed mindset to growth mindset:
  - Ask about how people like challenges, learn new things, stretch themselves, and see their current job as a step on their path.
- The presence and level of growth mindset discussions during training and/or employee performance reviews and appraisals:
  - Note: You will notice that 'during training, employee performance reviews and appraisals' and 'during leadership conversations' are starting to be recurring themes. This means that as you set up to talk to people about any of the elements of the *Essentials of Safety*, think about what others you can capture in the same conversation. It also highlights how interrelated they all are.
- The level of interest in learning as the purpose of investigations/learning studies amongst the individuals involved in the incident, the work team, and leaders:
  - Ask people 'Why do we do investigations?' 'What are investigations for?'
- The level of a culture that seeks to learn in the business:
  - How often is learning talked about by, and between, senior leaders in the business? How often do people hear 'okay, but what can we learn from this?'
- Views of any field leadership conversation activities and other leadership interactions, by those directly involved, and by leaders more generally:
  - This assumes that the business undertakes this type of activity. If not formally done, then explore any informal leadership interactions that occur in the workplace.
- The level of learning as a focus during Work-As-Normal on a day-to-day basis:
  - Does the business use or promote any kind of formal or informal learning such as Learning from Normal Work (LNW) reviews?
- Level of challenge in the role and the level of feeling and interest about the level of challenge in the role by those involved in the incident:
  - Some people are just happy to come to work, do their jobs, and go home. Others seek to excel and strive to challenge themselves. What is the balance here?

Has a High Level of Understanding and Curiosity about How Work Is Actually Done

- The level of drift in Work-As-Done during the incident, Work-As-Normal for the task on other days with other people, and the level of drift in the Work-As-Written.
- Variability in Work-As-Normal across the business for the task directly related to the incident.
- Size of the Work-As-Written system:
  - This refers to the number of procedures, work instructions, guidelines and standards, etc.
- Target audience and purpose of Work-As-Written 'safety' systems:
  - Some operations view procedures as tools to protect the backsides of the leaders, to keep them out of gaol. Others have a view that procedures are for following so we get work done the same way each and every time. Still others believe they are there to help those at risk – the workers – know what they need to get right in order to achieve safe work.
- Readability, useability, practicability, complicatedness, and complexity of Work-As-Written:
  - Specific to the incident and also more generally. This ties into the previous topic. For example, I have seen a working at height procedure, written for the end-user, run to 124 pages. This is ridiculous. Technically correct but useless as a day-to-day tool for doing work. Always ask the workers what they think of the procedures. Who wrote them? Who reviews them? What level of involvement by the end-user was undertaken during writing and review activities?
- Use and misuse of checklists:
  - Determine the level of tick and flick, or its digital equivalent click and flick.
- Task assignment effectiveness:
  - Always check both sides of this equation – those giving instructions and those getting instructions. The messages are not always the same. Also check both Work-As-Done and Work-As-Normal in this space.
- The process and frequency of supervisor checks during tasks:
  - How often do frontline supervisors visit the worksites and check in on those doing the work?
- Workload during the shift:

- This varies and is always subjective. Some people feel more pressure than others and try to get beyond 'It is always crazy busy here' to understand unusual workloads or production pressures and their regularity.
■ Numbers and details of procedures, standards, or work instructions required for the task:
- This is specific to the task involved. Think of an electrician doing some high-voltage live testing in a confined space whilst working at height with a boiler maker wielding an oxyacetylene gas axe next to them. Try to work out how many procedures and work instructions would be needed for that little task!
■ Levels of competence in the team for the task:
- Don't just check whether they have been trained. Explore how competency was determined, including the process for determining competency.
■ Details of any problems or difficulties with the task in the past, including the level of routineness or improvisation during the task:
- Both with the individual directly involved in the incident and also others who routinely do the work (Work-As-Normal).
■ Specific details of how the task was undertaken:
- This is essential for the Work-As-Done component of the timeline. Include here what stage of the procedure the team was up to when the incident happened.
■ Impact that other activities in the area may have on the task:
- Any SIMOPS going on? (SIMultaneous OPerations)
■ Confidence across the team that what they did was going to work:
- This also attends to the planning stage and the Task-Based-Risk-Assessment components of the task on the day.
■ Levels of attention needed when the task was being carried out:
- How critical and difficult were the steps?
- What was the level of Situational Awareness and on what?
■ Decision-making during the event (or when creating the Task-Based Risk Assessment):
- Who in the team are the decision-makers in the planning stages of the task normally, and on the day of the incident?
■ Details of team focus at the exact time of the event:
- Determine exactly what was going by each of the team members at the time.

■ The understanding of the situation (mental model) within the individual, the immediate work group, and within the leader of the work group:
  − How clear and consistent were people's views across the teams?

Understands Their Own and Others' Expectations

■ Expectations of the worker on the quality and outcomes of the work at hand.
■ Expectations of the leader on the quality and outcomes of the work at hand.
■ Expectations of the workers and leaders on drift and variability of Work-As-Normal:
  − Is drift and variability an accepted thing, or do workers and leaders view procedural compliance as a must?
■ Views of workers as to whether leaders' behaviours match their spoken expectations:
  − Is there a difference between the words leaders use and what actually goes on? − 'Procedural compliance is our number one focus'.
■ Balance between the words 'blame' and 'learn' after an incident:
  − Check this by talking with leaders, and importantly those who have been involved in previous incidents, along with other teams' views.
■ Behaviours associated with reading (or not reading) procedures:
  − You can usually only do this after exploring how many procedures and work instructions apply to a specific task.
■ Leaders' and individuals' views on 'Following the rules or doing the right thing?'
  − This is tied to the business's view on procedural compliance but goes deeper to explore the underlying beliefs.

Understands the Limitations and Use of Situational Awareness

■ The strong and weak signals that exist in the workplace on a normal day:
  − What do people need to keep an eye on in the general day-to-day kerfuffle of work, including what signals they are worried about?
■ Things that people keep an eye on during the task of interest.
■ What occurred related to the incident that the individual or work team were surprised about during the incident?

■ Details of the shared mental models that existed between individuals in the work team (Work-As-Normal and Work-As-Done).
■ The existence and efficacy of routines such as task post-mortems and what they turn up:
- Are there any examples of post-work review activities across the business?
■ Whether there are any routine surprises associated with the task related to the incident.
■ Details of things that appeared or that happened during the incident that were not predicted during the Task-Based-Risk-Assessment:
- Include what the work crew normally does when unexpected things pop up.
■ What assumptions are we making about how the work is done as compared to how we thought it was when we wrote the procedure?
- You will need to have a conversation with the creator of the procedures and work instructions associated with the task to answer this and other questions about intent, context, etc.
■ The next few are related and explored through the specific incident-related work team conversations:
- Do all know their responsibilities and activities for this job?
- What do they think will go right? What will go wrong?
- Is the work team seeing any drift associated with situational awareness? Are there any changes in what they need to keep an eye on whilst doing the task, compared to how they have done it in the past?
- What happens normally that may influence their ability to keep an eye on what they need to in order to get the job done?
- What is the work crew process when something unexpected happens? For example, an interruption, a new and urgent task, an unexpected change of conditions, or a resource that is missing.
■ Process monitoring activities during the task:
- How often, how, and by whom is the progress, quality, etc. of the job monitored?
■ What are the observation activities for things that could become a threat in the near future?
■ An exploration of the observations prior to or during the task that triggered thoughts that something might not go as planned:
- Do peer-on-peer observations take place by teams during tasks? If so, what do they look like? How are they viewed?

Listens Generously

- Listening levels between a direct leader and workers related to task allocation and task acceptance/understanding:
  - What are the views of the workers about how well their leaders listen to them as they work through the task allocation and expectation discussions?
- The level of listening skills at various levels of leadership in the business:
  - This is usually a judgment call based on the numerous conversations held as a part of the learning study rather than a specific set of conversations around listening. You should easily pick up a leader's listening skills by having a chat with them about the incident more generally.

Plans Work Using Risk Intelligence

- Level of planning and details of planning that went into the incident-related task and Work-As-Normal:
  - Check out both the low-level, on-the-day, by-the-crew planning for the task as well as the broader planning activities related to it.
- ETTO balance (time to think – time to do, blue work – red work):
  - Especially by those who planned the task, what is the understanding of not getting 100% of everything 100% accurate 100% of the time and spending sufficient time thinking about things, and how does that play a part in their leadership and planning activities?
- Individual and work team expectations regarding likelihood of success and likelihood of failure prior to starting the work:
  - This is focussed on what they thought as they were pulling the activity together rather than a more general discussion on the likelihood of success.
- Level of chronic unease in the team and the leadership teams:
  - How did this wariness for the effectiveness of controls manifest in conversations between individual team members, and between team members and leaders during normal day-to-day activities?
  - What was the level of preoccupation with failure amongst leaders, amongst work teams, amongst individuals, and amongst those directly involved in the incident?
- Level of wariness of the effectiveness of the controls:

- By the workers from the impacted area and from other areas, as well as amongst functional staff and leaders.
■ Knowledge of any previous events or near-misses involving this task or similar tasks.

Controls Risk

■ Levels of hazard identification in the planning stages of the task related to the incident and during Work-As-Normal:
  - This is very much about the quality of any HAZOPs, HAZIDs, and also any last-minute risk assessments such as Task-Based-Risk-Assessments, Job Safety Analyses, Task Hazard Analyses, etc. Also consider the general quality of those risk assessment activities. Include who gets involved in the creation of these risk and hazard assessments.
■ Level of consideration of the hierarchy of controls, line of fire, and SIMOPS considered in the Task-Based-Risk-Assessment:
  - Check this out as directly related to the incident (Work-As-Done) and also as Work-As-Normal.
■ Understanding and treatment of trigger steps generally and in the task related to the incident.
■ Understanding and treatment of critical steps generally and in the task related to the incident.
■ What is the balance between 'framework' sections and 'critical' sections within procedures and how are they treated?
■ Existence of any anomalies in the workplace during the development of the incident that do not appear to have been there before:
  - Did anything stand out as different or unexpected?
■ What devices, alarms, and warnings were provided or used immediately prior to the incident, and how effective were they?
  - What were the options considered for foreseen failures in the task and escape plans?
■ Level of use by the incident involved team and/or the business of tele-graphing deliberate action.
■ The number of safety signs in the workplace, including people's awareness and thoughts about them:
  - This needs a good walk around and simple short chats with people in the workplace.
■ Level of controls – shared space – over protection:

- This needs to be from the perspective of both those who do the work and those who lead them.
■ Use of Critical Control Verification leadership activities:
  - Check out both formal and informal critical control verification activities here.
■ Human Factors/Ergonomics (whether the workplace contributes to or hinders safe work).
■ Details of what tools and equipment are normally used to get this task done:
■ Incident related and any variance with Work-As-Normal tool and equipment selection and use.

Applies a Non-Directive Coaching Style to Interactions

■ Level of coaching:
  - Is there a formal process?
  - Is there a coach-the-coach process?
■ Use of a coaching style in leaders' behaviours:
  - Also considered in the section on leadership, how do the leaders engage with their people with respect to a coaching style of management and leadership?
■ Quality of the feedback process:
  - Check out both formal and informal feedback mechanisms in the business.

Coaching and the use of a coaching style in leading and managing can play a direct and indirect part in incident formation as it relates to and informs more general leadership inputs to the workplace and the incident.

Has a Resilient Performance Approach to Systems Development

■ Level of resilience in the systems, procedures, and leadership behaviours:
  - Respond: Knowing what to do when things start moving away from going right.
  - Monitor: Knowing what to look for or being able to monitor things that need to be in place to ensure things go right.

- Learn: Knowing what has happened and being able to learn from the experience.
- Anticipate: Knowing what to expect or being able to anticipate developments into the future.

■ Safety oscillation related topics:
- How does the management modify their focus when safety results go down?
- How does the management modify their focus when safety results go up?
- How does the management modify their focus on safety when production goes up?
- How does the management modify their focus on safety when production goes down?
- How does the management modify their focus on safety when the business budgets are cut?
- Use the model here to help formulate questions to ask and data to be collected in relation to the Safety Oscillation Model.

Adopts an Authentic Leadership Approach When Leading Others

■ Balance of tell and ask amongst leaders:
- Observed during discussions with workers.
■ Level of intent-based, adaptive, and/or authentic leadership:
- This can be sometimes difficult to get at during a simple learning study activity. Consider asking for any leadership reviews or audits that have taken place in relation to the measurement activities described in Chapter 5, and get your head around those reports.
■ Use of a coaching style in leaders' behaviours.
■ Review of leadership development plan detail. Does it include all of, none of, some of:
- Inquisitive mindset.
- Listening.
- Acceptance of being proven wrong as well as be proven right.
- Chronic unease.
- Decision-making.
- Intent-based leadership.

– Focus on achieving excellence compared to preventing failure.
– Deference to expertise.
– Seeing failure as learning.
– Asking compared to telling.
– Coaching style.
– Providing soon, certain, and positive feedback.
– Adaptive problem-solving.
– The *Essentials of Safety*.
■ Leadership *to* the people or *with* the people:
– This must be from the perspectives of both leaders and followers.

Example: Using the Essentials of Safety in a Workplace-Incident Learning Study

From the start, let's ditch the term 'investigation'. Even calling it an 'investigation' is fraught with debate about 'root' cause, who is accountable, and the danger of losing the idea that this is all about learning. Various authors and safety thinkers, especially Todd Conklin, talk about 'learning teams' as a surrogate for the word 'investigation', which works well in the way he describes and practices it. I will use the term 'learning study' for what I am describing here as it has some elements of making sure we understand how the real world works and what we can learn from it, and also reminds us that we have to do a study. We have to do some exploration and thinking work here, and that the process does need some leadership, just like any study does. A learning study can range in complexity from a simple study looking at Work-As-Normal (see the 'Learning from Normal Work' review process at the end of this chapter) to highly complex and fluid incidents which may seem really hard to get our heads around.

There is one guarantee in the workplace and that is that people will do things we do not expect them to. In this example, we explore, using the framework of the *Essentials of Safety*, a method for undertaking workplace-incident learning studies. Not surprisingly perhaps, it pulls heavily on my previous book *Simplicity in Safety Investigations* inasmuch as it talks to the imperative in any investigation method to understand how the work was done (Work-As-Done), how it was normally done (Work-As-Normal), in addition to how we think it is done (Work-As-Written). We use the *Essentials of Safety* elements to explore the fundamental question in a learning study: 'What played a part in creating the incident and what can we learn from it?'

A word on our mindset before undertaking a learning study after a workplace safety incident. What we look for, we find. If we are truly curious as to what happened and what we can learn from it, then we will think, talk, and behave in alignment with that mindset. This is where our minds need to be as we form the team to explore what did not go quite as we anticipated it would.

Also, before launching into the details of the learning study process we need to understand the learning study basic steps:

1. A decision to undertake a learning study is made.
2. A preliminary information-gathering step is undertaken.
3. A learning study team is formed.
4. The team goes through the draft Work-As-Done timeline.
5. More comprehensive information-gathering is undertaken by the team, including a PEEPO (People, Environment, Equipment, Procedures, and Organisation).
6. The timeline is completed and 'Elements of Interest' identified. These are the gaps between Work-As-Done and Work-As-Normal and/or the gaps identified between Work-As-Normal and Work-As-Written.
7. The contributors are listed for each Element of Interest.
8. For each sentence, word, or statement created within the contributors, the team explores what we can learn and how we may embed those lessons.
9. Capture the timeline, gaps, exploration, and learning into a simple report.

Let's have a look in detail at each of these steps:

1. A decision to undertake a learning study is made
A learning study can be undertaken on just about anything. It could range from a safety, health, occupational hygiene, or environmental incident, a production or financial loss, or a failure in a health system. The incident could also be a non-incident. By this, I mean that we can undertake a learning study when things are going well in addition to when they are not going so well. We call this an LNW review. The LNW review is covered in more detail at the end of the chapter.

A decision as to whether to undertake a learning study needs to be made based on the actual and potential severity of the incident, whether the incident or one similar to it has occurred in the past, and whether there appear to be lessons to be learnt by the organisation. About the worst thing you can do for the quality of learning studies is to run them based on a number derived from key performance indices (KPI's). I have seen businesses that require x investigations to be undertaken per month. This is not a good behaviour in my opinion and devalues the investigation processes. There will most likely be some local legislative requirement that requires you to carry out an investigation after a workplace incident. The legislation will rarely specify what sort of investigation to do, only that one needs to be undertaken. A good approach is for the leader of the area to pause, talk with peers and advisors, and think about what happened. They need to contemplate whether there are things that we might learn from the incident and what level of resources we need to spend on it. A simple example I saw in a webinar recently was in relation to two incidents. One involved a truck operator falling about 1.5 metre off a truck access ladder when a handrail broke. The other was about a guy tripping over his own feet on a flat surface. Both resulted in medical treatment and a legislated need to investigate. The question to consider is how much effort you are going to put into the truck incident learning study and how much effort will you put into the 'tripped over his own feet' incident. Without knowing any more specifics, it appears that there could be more to learn from the truck incident and so it should command a higher level, broader, more powerful examination than the 'tripped over his own feet' incident. You can also use a reflection of the 'thinking' elements of *Essentials of Safety* to inform your decision to seek to learn from the incident. For example, a difference between Work-As-Done and Work-As-Written may be triggered because of people's cognitive limits being overwhelmed and you want to see what we can learn from this.

2. A preliminary information-gathering step is undertaken

The focus of the preliminary information-gathering step is on what occurred (Work-As-Done). There is usually also a smattering of what the Work-As-Written was, and some Work-As-Normal where its need is obvious – it is always a good idea to talk with others who do the same work as that involved in the incident about Work-As-Normal as early data and information are collected.

This initial information-gathering activity is undertaken by the local area as soon as possible after the incident and would typically involve a balance between post-incident conversations (interviews) and some document and system digging. The output of this step usually consists of a preliminary, or draft, timeline (Work-As-Done only) and an armful of interview notes, procedures, task-based risk assessments, training records, and other related documents and notes. It is important to ensure that some early information gathering is done, especially concerning interviews and conversations as the human memory is fallible and erodes very quickly over time.

3. A learning team study team is formed

The size and level of horsepower of the team will depend on the incident. The more complex or complicated the incident appears to be and the more there may be to learn from the incident, the larger and more highly powered the team should be.

There are a couple of mandated positions within any learning study team that you should never be without:

- An independent learning study team leader. The level of this leader should be equivalent to the owner of the incident and in alignment with the depth of the study. For example, if the incident was a significant, life-changing injury, which could easily have been a fatality, the independent leader should be no less than a manager (or their equivalent). If the incident was a simple rolled ankle and there were a few things to be learnt for the business, a superintendent or equivalent may be perfect for the task. The reason we need an appropriately senior person in the team leader role is that they generally have a good understanding of how an organisation is designed and functions at a strategic level and this is important when understanding the organisational component of the incident.
- A learning study process facilitator. Make sure they know how to run a learning study and clearly display the skill of herding cats. A learning study, like any other investigation process, can easily go off track if not facilitated well.
- One or more subject matter experts (SME). People who know the ins and outs of the systems, and the technical processes that lie behind the tasks being undertaken at the time of the incident. This could be the technical engineering type people who know the process back to front and inside out, or the specialist nurse or medical practitioner who has been around long enough to know what is going on and is recognised as such by managers and peers.
- A number of 'real people'. These are people who do the job that was being done at the time of the incident. They can include peers from other shifts or other parts of the business. Pick people who will happily share how things really are out there with the team. I believe that there needs to be two real people as team members. These are the source of most of the Work-As-Normal information that is essential to understanding in the learning study.

■ An alternative thinker. This is an important role, especially for complex, technically specific incidents. This team member brings the challenging questions and the upsetting of the status quo of the SMEs and should ideally not know too much about the technical components of the task related to the incident. Another way to view this role is as the black hat of DeBono's *Six Thinking Hats* fame.

The total number of people making up a learning study team will depend not only on the complicatedness or complexity of the incident but also on the skill of the facilitator. Facilitating any process with 15 or so highly motivated and sometimes boisterous people with strong opinions in one room at one time is not a task for the faint hearted – I know this from personal experience. I have found that the most effective and efficient number of team members is in the range seven to nine.

4. The team goes through the draft Work-As-Done timeline
Prior to this step, it is useful to give an overview of the incident and of the learning study process. Considering that only a preliminary set of information has been collected at this point in time, the timeline that is given to the team will most likely not contain all the Work-As-Done details that the team will need for the learning study. Nor will it contain any of the Work-As-Normal and Work-As-Written detail required. The purpose of this step is to give the team a bit more information concerning the incident and to prime their minds for the next step.

5. A more comprehensive information-gathering activity is undertaken by the team
To work out what additional information may be required for the Work-As-Done, Work-As-Normal, and Work-As-Written parts of the timeline and also for the exploration of any gaps, we carry out a more formal PEEPO. PEEPO is simply an acronym for People, Environment, Equipment, Procedures, and Organisation. It is a simple process that enables the team to think about and record the data and information they want to be collected to enable them to understand the incident.

Carrying out a PEEPO is one of the simplest tasks of the facilitator. You need a white board and some small sticky notes (3M brand A7 sized PostIt® notes are the best).

Disperse the sticky notes amongst the team and ask them to write down anything they would like to see in the way of data under any of the PEEPO categories. A great way to encourage them is to ask questions such as 'Okay Jim, you are a Safety Rep (or electrician, or whatever). What is it that you would like to see that would help you understand exactly what happened here?' Get them to write down the data they would like to see collected and then stick the note up on the white board under PEEP or O.

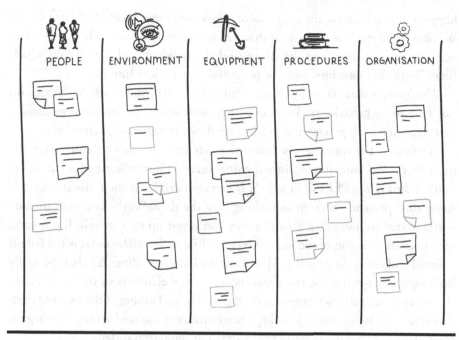

| PEOPLE | ENVIRONMENT | EQUIPMENT | PROCEDURES | ORGANISATION |

Figure 4.1 A completed PEEPO.

Additional matter for consideration as the PEEPO is undertaken relates to the *Essentials of Safety* discussed in the main section of this chapter above.

Once the PEEPO is completed (this normally takes 45 minutes or so), distribute the sticky notes to the team (except the facilitator) and instruct the team that they are accountable for collecting the data and bringing them back with them when the team reconvenes. Set a time with the learning study team leader for the next session based on the complexity and time constraints for the collection of the information from the PEEPO.

The time it will take for the data to be collected will vary enormously. Before the team gets back together after collecting their information, ring around and make sure all of the information has been collected. If not, postpone the next section of the learning study and offer assistance in the gathering of the information. It is not a good idea to finalise the timeline development and the rest of the learning study with only a portion of the PEEPO data collected. In fact, generally, from my experience, it is a terrible idea (Figure 4.1).

6. The timeline is completed and 'Elements of Interest' identified
The timeline is the mechanism for aligning the learning study team on 'what' happened during the event, what happens normally, and what 'should' have

happened. It is built on the concepts of Work-As-Done, Work-As-Normal, and Work-As-Written. The purpose of this step is also to spend some time, as a team, getting the team members' heads around the incident and identifying which (few) Work-As-Done bits could be potential Elements of Interest.

Don't forget that Work-As-Done, Work-As-Normal, and Work-As-Written do not have to be behaviours. They could represent system issues, instrumentation variances, financial, process and production flows, maintenance activities, etc.

To finalise the timeline, the learning study team focuses on converting the draft Work-As-Done timeline into the final version after collecting further information during the PEEPO in step 5. It is most effective if the facilitator spends some time prior to this step in making sure the draft Work-As-Done timeline is transferred to individual PostIt® notes and is put up on the wall. It is a good idea to cover a section of the wall with paper first as the sticky notes often fall off painted and often dusty walls. The Work-As-Done timeline can then be easily built upon by getting the team members to write elements of the timeline on the sticky notes and stick them up on the wall in and amongst the existing draft timeline. The facilitator will be kept busy ensuring the sticky note language is clear and the sticky notes are in the correct chronological order.

Particular care needs to be taken with the wording of the Work-As-Done notes in order to facilitate the creation of Work-As-Normal and Work-As-Written. For example, you will need to couch negatives into an element of the Work-As-Done so that it sticks out as something worth exploring. For example,

Insufficient information	Better clarity of the issue
The supervisor gave an instruction to swing valve 24.	The supervisor gave an instruction to swing valve 24 without discussing the associated hazards and controls.
The scaffolder climbed off the scaffold at height.	The scaffolder climbed off the scaffold at height without wearing a harness and lanyard.
The carpenter used a hammer.	The carpenter used a hammer to insert a screw into the table top.

As you can see, the descriptions in the 'Better clarity of the issue' column lend themselves to the building of Work-As-Normal and Work-As-Written more easily than those in the 'Insufficient information' column.

Once the Work-As-Done portion of the timeline is complete and all team members are happy with it as a description of what happened, they need to decide which Work-As-Done elements need to be completed into an 'Element of

Interest' by the addition of Work-As-Normal and Work-As-Written. I often use a simple voting system to decide which elements of the Work-As-Done timeline are going to be built into Elements of Interest. Giving each team member three to five ticks each usually works as a voting method. Once the team has identified a (small) number of Work-As-Done elements that they want to build on, the team adds a sticky note above the relevant Work-As-Done describing the Work-As-Written associated with it. Then they add the associated Work-As-Normal. The topics must be clearly the same. The team will need to pull strongly from the data collected by them for this. For example,

Element	Poor element of interest	Better element of interest
WAW	Supervisors should discuss hazards with operators.	Procedure 24 – Task Assignment – requires that the hazards and controls associated with a task are discussed between the person assigning the task and the person doing the task during task assignment.
WAN	Supervisors often give instructions to swing valves over the radio.	Supervisors often assign tasks without discussing the associated hazards and controls for the task.
WAD	The supervisor gave an instruction to swing valve 24 to the operator.	The supervisor gave an instruction to swing valve 24 without discussing the associated hazards and controls.

Element	Poor element of interest	Better element of interest
WAW	Carpenters must use a screwdriver to put screws in.	Work instruction 24 – Screw Insertion – requires carpenters to use a screwdriver to put screws into tabletops.
WAN	It is reasonable to expect carpenters to use screwdrivers.	It is common practice for carpenters to use hammers to insert a screw in tabletops.
WAD	The carpenter used a hammer.	The carpenter used a hammer to insert a screw into the tabletop.

The facilitator then circles the gaps. In both of the above examples the gap lies between Work-As-Normal and Work-As-Written. You should aim to have no more than four or five Elements of Interest. Any more than that and you are probably looking at elements that are not in the causal pathway of the event and may need to be captured in a parking lot instead.

At this time, we have a completed timeline with the identified Elements of Interest (Figure 4.2).

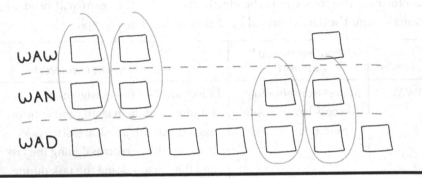

Figure 4.2   A complete timeline with gaps identified.

7. For each of the gaps identified between Work-As-Done and Work-As-Normal and for each of the gaps identified between Work-As-Normal and Work-As-Written (Elements of Interest), capture the contributors from the information collected

Before getting into the activities of this step, it is important for the team to have sufficient soak time so that the team can establish a common mental model of the incident, where the gaps are, what the Elements of Interest are, and an over-view of the information that has been collected by the team. The time needed for this can vary a lot. Take the time needed so that the majority of team members and happy to move onto this step.

Identifying the contributors is a white board/brainstorm activity with the whole team and covers each Element of Interest and sometimes the incident itself. The first part of this is the secret to a good process outcome. We need to ask the right question to start with. Each question is a translation of one of the Element of Interest gaps into a 'Why' question. Following on from this, the team pulls from the data it has collected and builds a set of statements, words, and sentences that describe the contributors. These are then captured on a whiteboard. A useful prompt to help the study team can be things like:

■ What information have we collected during our PEEPO that relates to this Element of Interest question?

- What Essentials of Safety element or part of the total system let us down here?
  - Have we gone deep enough to understand this as yet?
  - Do we need to ask another question to go deeper here?
- What else?

Here are a couple of examples from a scenario loosely based on the *Herald of Free Enterprise* incident. I say loosely based as I have taken the original report, books, etc. and tweaked them so that the incident and its details are more useful as a learning tool for when I run learning study training sessions. I encourage you to read the full report if you want to get a better picture of the actualities of the incident. In a nutshell, the *Herald of Free Enterprise* incident is described as:

On 6 March 1987, the Roll on/Roll off passenger and freight ferry *Herald Of Free Enterprise* under the command of Captain David Lewry sailed from Number 12 berth in the inner harbour at Zeebrugge at 18.05 G.M.T. The bow doors were left open after loading vehicles and passengers. The *Herald* was manned by a crew of 80 hands all told and was laden with 81 cars, 47 freight vehicles, and 3 other vehicles. Approximately 459 passengers had embarked for the voyage to Dover, which they expected to be completed without incident in the prevailing good weather. There was a light easterly breeze and very little sea or swell.

The *Herald* passed the outer mole (harbour marker) at 18.24. Massive volumes of water entered through the open bow doors and the *Herald* capsized about 4 minutes later. During the final moments, the *Herald* turned rapidly to starboard and was prevented from sinking totally by reason only that her port side took the ground in shallow water. The *Herald* came to rest with her starboard side above the surface. Water rapidly filled the ship below the surface level with the result that not less than 150 passengers and 38 members of the crew lost their lives. Many others were injured.

The first gap identified was between Work-As-Normal and Work-As-Written. The Element of Interest looked like this:

WAW – *General Instruction (1984) requires that the loading officer ensures the bow doors were 'secured when leaving port'.*

WAN – *It is routine for the loading officers across the Free Enterprise vessels to leave for their harbour stations without ensuring the bow doors are 'secured when leaving port'.*

WAD – *The loading officer (Chief Officer) left the bow door area without ensuring the bow doors were 'secured when leaving port'.*

The Gap question therefore is between Work-As-Normal and Work-As-Written. The question becomes, therefore:

'*Why is it common for loading officers across the Free Enterprise vessels to leave for their harbour stations without ensuring the bow doors are 'secured when leaving port'?*'

An example list of contributors for this Element of Interest could look like this:

- The Masters of Free Enterprise vessels rely on a negative reporting process, which relies on people telling him or her when something is NOT right.
- There are conflicting and ambiguous TCF Limited standing orders and procedures, especially related to required activities prior to leaving port.
- There are fewer officers on the Dover – Zeebrugge run than other TCF runs.
- The loading officers were routinely not queried by the Master about the open or shut status of bow doors.
- The closing of the bow doors was not deemed as a 'critical' step in the process of leaving port.
- There was no business-wide learning from five previous TCF 'sailing with bow doors open' incidents.
- There was pressure to sail without verification of critical pre-sailing steps (e.g. shutting the bow doors).

After the first Element of Interest's contributors are captured on the whiteboard, write them up on a large butcher's paper and stick it up on a wall. Repeat the process until you have explored all of the Elements of Interest identified.

Another example of a gap identified was also between Work-As-Normal and Work-As-Written. The Element of Interest looked like this:

WAW – *Section 68(2) of the Act requires the Master to know the draught of their vessels prior to putting to sea.*

WAN – *It is routine for Free Enterprise vessels to sail down at the head without checking the draught of the vessel.*

WAD – *The Master gave the order to sail without ensuring the fore and aft draughts were checked.*

The gap identified was between Work-As-Normal and Work-As-Written. The gap question therefore is:

Why is it routine for Free Enterprise vessels to leave Zeebrugge port trimmed by the head?

The white board gap contribution brainstorm revealed the following thoughts:

- There was no instrument available to measure the draught of the vessel.
- The TCF senior leaders did not listen to questions and concerns raised by Masters.
- The ballast 14 pump was not sufficient to timeously empty tanks whilst at the port.
- There was limited to no awareness of trim characteristics of Free Enterprise vessels by Master or TCF management.
- There was no level of chronic unease with respect to sailing trimmed by the head. (No what if?)
- The trim of the vessel not considered as a critical control.
- TCF knew it was not possible to comply with draught measuring legislation and did nothing about helping the Masters comply.

The output of this step is two or three sheets of paper containing the contributors to two or three Elements of Interest.

8. For each sentence, word, or statement, explore what we can learn and how we may embed those lessons

For each of the Elements of Interests, we potentially now have a piece of butcher's paper full of statements, words, and sentences that describe the contributors to each Element of Interest and therefore to the incident. These need to be sorted and converted into lessons that the business can learn from.

There will be a number of lists and that may describe too many contributors to effectively manage the lessons for. In order to make the lists more manageable, they can be consolidated into categories if needed.

The categories should make sense for the specific incident and your business. A couple of examples of category lists are:

- Systems.
- Risk Control.
- Leadership.

Or maybe:

- Plant.
- Process.
- People.
- Organisation.

Under each of these topics, we place the contributors that fit into that classification regardless of which Element of Interest they came from. We then need to ask:

- What can we learn as an organisation, and as individuals, from this incident?
- How can we embed that learning into our systems, behaviours, beliefs, routines, and ways of being in our business?

There could be a number of lessons from each category. Whilst this appears to be a duplication of work as we move items from the first Element of Interest focussed list to the category list, it is a very helpful activity and actually makes the process flow more effectively. We need to identify the lesson itself and what the action is to promote the sustainability of the learning from the lesson.

By way of example, again from the mock-up of the *Herald of Free Enterprise* study used in the training, the 'Risk Control' list was as follows:

Risk Control:

- There is a lack of aligned understanding of speed characteristics of Free Enterprise vessels.
- There was no level of chronic unease with respect to sailing trimmed by the head. (No 'what if?' was ever asked.)
- There is limit to no awareness of trim characteristics of Free Enterprise vessels by Masters of Free Enterprise vessels or TCF management.
- The ballast 14 pump is not sufficient to timeously empty tanks whilst at the port.
- There was no instrument available to measure the draught of the vessel.
- No alarm or instrument existed on the bridge that could alert officers of the open or closed status of the bow doors.
- The loading officers were never queried by the Master about the bow doors being shut prior to leaving port.

Derived from this list is the first lesson, from the category 'Risk Control'.

Lesson 1: 'Operators of complex equipment such as ferries need to be familiar with the operational characteristics and limitations of the equipment they operate in order to be able to manage and operate within those characteristics.'

We then explore what action or actions we might take in order to sustain the learning from lesson 1 and embed it into the systems, behaviours, beliefs, routines, and ways of being in our business. A sustaining action for this lesson could look like: 'Include in the training and competency assessment systems for operators (masters, chief officers, senior engineers, and senior decision-makers) a requirement that they study and deeply understand the equipment they are expected to manage and operate. Especially where those characteristics relate to the safe operation of that equipment.'

By way of another example, again from the mock-up of the *Herald of Free Enterprise* study used in the training, but this time from the 'Systems' list:

Systems:

- There is no system for consideration and management of 'critical' tasks.
- The trim of the vessel was not considered as 'critical'.
- The closing of the bow doors was not deemed as a 'critical' step within the 'leaving port' process.
- There was no learning from five previous 'sailing with bow doors open' incidents.
- There are conflicting and ambiguous TCF limited standing orders and procedures.
- There is a masters' negative reporting process which relies on people telling him when something is NOT right.

The second lesson is then derived from the contributors in the category 'Systems' as outlined above. One could look like this for example: Lesson 2 – 'Processes must be in place to ensure that lessons from incidents are shared in a way that results in changes in the way people believe, behave and the way they control risks.'

A possible action to sustain the learning from lesson 2 could be along the lines of: 'Implement an incident learning management approach that ensures that learning from incidents are shared in a way that results in changes in the way people behave and control risks. This must not be limited to leadership verification processes, training, procedures, standing orders, reporting processes, or critical controls'. In both examples, other lessons could also be captured. I have chosen the ones I did purely as examples.

9. Capture the timeline, gaps, exploration, and learning into a simple report
The secret to a good-quality workplace incident report is to keep it simple. It should contain as a maximum:

- Title of the incident.
- Executive summary.
- Photos.
- Learning Study team members.
- Timeline, including the Elements of Interest highlighted or circled.
- Contributors by topic list (e.g. 'Systems') or by Element of Interest.
- Lessons to be learnt.
- Actions to sustain the learning from the lessons.

Even if you do not adopt the full approach, consider using this example as a way of thinking about investigations. Explore what are the contributors to workplace incidents and what can we learn from them.

Chapter appendix A: Post-incident data-gathering conversations (AKA Interviews)

Historically, these where called 'interviews'. To me, they are simply data-gathering conversations or a 'chat with a purpose'.

Much of this bit is from *Simplicity in Safety Investigations* and also *Investigative Interviewing* by Rebecca Milne and Ray Bull.

Before getting into the details of what to say and how to say it, remember what we are there for – to better understand a person's story about a series of events. We are not there to gather information owed to us. We are not there to gather 'evidence'. We need to set up our minds to listen generously. And we need to set up the conversation so that it happens as soon as possible after the incident. Memories fade and are altered far more quickly and permanently than we often realise.

As we have previously talked about, generous listening is an art. It is an art that needs constant attention and practice. It is the art of using your whole body and self in the act of listening. It is about listening to understand rather than listening to respond. It is about paying attention. It is not about the person doing the listening – it is about the person being listened to. The topics and themes being discussed are very important. However, the way in which we have these conversations is more important.

Think of the process of gathering information from those involved in a workplace event as a conversation rather than a list of questions that must be answered or as a formal interview. That said, you will of course need to ask some questions. It is how you ask the questions that makes the difference. I have found the list of *Essentials of Safety* element topics and ideas very helpful with post-incident conversations in the past. I use a prepared list each and every time I have a conversation as a part of information gathering after a workplace incident. I do not read all of them, but I do scan the prompts and select some key topics and ideas that are relevant to the incident being investigated. It helps get my mind in the right space for the conversations I'm about to have.

Remember, you are trying to understand a person's story around a workplace incident that you can either talk or you can listen. Do not try to do both at once – it is not possible. The next time you are having a conversation with a friend and they are telling you something, start talking to them about something else and see if you can still listen to them *and* keep track of what your friend is talking about. It is not possible – unless you have the skills of a simultaneous translator.

If you are having a conversation with someone in order to find out more about an incident and you interrupt them, you have done a couple of major things wrong. One is that you have stopped them talking and hence you have stopped them thinking. You have also interrupted your own flow of listening, and as I have said before, it is not possible to listen and talk at the same time, so just be quiet and listen.

In summary then, there are a number of things that we should try to focus on when listening and talking during a post-incident conversation. They are:

■ Listen, and stop talking.
■ Be prepared for the conversation. Go through a prompt list extracted from the expansive list earlier in the chapter. Think about the topics you would like to hear more about during the conversation. Try to put other distractions out of your mind.
■ Prepare the person you are having the conversation with by helping them be at ease. Remind them that this is just an information-gathering activity, similar to a meeting and not some formal Human Resources or legal process.
■ Take enough notes.
■ Be truly empathetic to the feelings of the person you are talking with. They may be nervous and uncertain. Try to stand in their shoes and see the issues and actions from their perspective.

- Be mindful that they may be upset and behave accordingly.
- Use your body language to support the person. Cues are useful here so nod or use short words of encouragement. Stay focussed on the person. Maintain appropriate eye contact.
- Be patient. Do not argue. Silence is a great tool to use. Let the person take as much time as they need to tell their story.
- Remove distractions from the environment. Turn your mobile phone off. Don't shuffle papers. Don't check email and don't keep looking at whoever is walking past or your watch.
- Never be judgmental. Simply listen to understand.
- Listen beyond the words. Listen to the tone and the alignment between the words and body language. Listen for what is not being said as well as what is being said.

In general, memory is pretty stable for the first couple of hours after an incident with more memory loss after that. So, if you cannot have the conversation within the first 2 hours, leaving it a bit is not always a problem.

The following phases are by no means a procedure that you must follow, they are simply intended to provide you with different techniques and ways of approaching the conversation that may help yield detailed information that will help with your investigation or learning study. One of the most important phases, that you should endeavour to do each time, is phase 1 – greet and build rapport. This will help set you up for success.

Phase 1: Greet and build rapport

It is essential for the people involved in the conversation to be as relaxed as possible. A part of this is building rapport and being comfortable in the conversation. It is a good idea to start with neutral questions that are not related to the incident itself. You also need to communicate empathy and an understanding of the situation from the other person's perspective.

Active, or generous. listening is all about listening with intent and for a purpose. The purpose is to listen to understand. It is not easy and requires you to maintain concentration. Listening is not just about the ears either. During the conversation, non-verbal behaviour (e.g. body position, hand signals) can also be important. By speaking slowly in a calm, even voice and behaving in a relaxed way can help calm the other person also. Avoiding hectic arm movements and speaking in a soft tone of voice can promote a relaxed atmosphere in the interview. Allowing time for people to give an elaborate answer and using pauses can help encourage a thorough search of memory. You can promote extensive answers during these pauses by remaining silent or by expressing simple utterances conveying their expectation that the person should carry on (e.g. 'mm-hmm'). This non-verbal behavioural feedback should not be qualitative (e.g. saying 'right' or 'Yes, I agree'). Repeated interruptions soon teach people that they have only a limited time to reply, which often leads to shortened responses to future questions.

Phase 2: Cover the purpose of the conversation

Make sure that you explain why you are having the conversation – in order to understand the story and to learn. Encourage and highlight the importance of reporting all information – including partial information, even if it seems unimportant. Remind people not to fabricate or guess what happened. If they are uncertain, don't encourage them to answer, preferably they should indicate if they 'don't know'. Transfer the control of the conversation to the other person by actively listening and building questions from their answers.

Phase 3: Initiating a free recall of the event

The secret here is to let the other person just talk. Ask the person to mentally relive the environmental and personal set of circumstances or facts that surround the particular event – encourage them to think about the sights, sounds, their feelings, and their emotions when the incident occurred. Encourage them to close their eyes when recalling the event. Pause to enable plenty of time to respond and conduct a thorough search of memory, and ask fewer but open-ended questions, encouraging them to simply talk.

Phase 4: Questioning

The way we ask questions can have a direct bearing on the information we get from the questions. The use of open and closed questions at the appropriate time is critical here. Asking mainly open-ended questions will encourage the person to talk more. Closed questions can be important but use them only when you need clarity about a topic. This is about tailoring your questions rather than having a standard set of questions or checklists. Do not ask leading questions, ever.

Phase 5: Varied retrieval

I find having a white board or a piece of paper helpful in getting good information. Some people prefer pictures to words. I always have photos of the incident scene at hand and often undertake the conversations out in the field where the incident occurred – you may not want to do this in the case of a fatality. Sometimes it is useful to suggest the person tells the story by working backwards from the end to the beginning instead of the other way around. I always start by exploring their Work-As-Normal, not the incident itself.

Phase 6: Summary

Repeat in summary form, the person's account of the incident. Encourage them to check for accuracy and also add additional information if they wish.

Phase 7: Closure

The idea is to attempt to finish the conversation with the person in a positive frame of mind. Go back to neutral topics slowly and aim to leave them with a positive last impression.

Chapter appendix B: Learning from getting it right: How to run a simple LNW review

LNW reviews can be undertaken at any point in time. It could be as a result of a series of low-level incidents that are appearing as a trend that you want to understand or from an interest arising from incidents in other parts of the business or industry. They are also initiated from field leadership interaction activities within your own business. The example I will talk about is based on this field leadership interaction process – whether you have a formal or an informal leadership interaction process for this activity or practice does not matter. I will cover off what one of these is now, so if you already have an embedded system just skim over this bit. The idea for a field leadership interaction process is that leaders interact with workers in the location where the work is being done. Basically, field leadership is where a leader visits a worker and discusses the work, the risk controls, critical steps and trigger steps, etc. so that the leader and the worker can walk away with a shared mental model of the work. It is also useful in sharing good practices across the workplace.

We have tended in the past not to do investigations when nothing has gone wrong. And we have not used a positive leadership interaction in the field as the basis for an investigation. This is exactly what I am proposing here.

I believe strongly that leadership conversations before an incident should be the same as leadership conversations after an incident – attempting to understand differences between how the work is being done and how we think it is being done.

I have spoken about the importance of the need to understand how work is normally done as being an essential component of doing a learning study (or any other type of investigation). Extending this idea into our normal day-to-day activities as a leader is where I want to focus.

Overall, a field leadership programme is all about getting our leaders out into the field and having conversations, undertaking verifications, observations, and engaging activities for the purposes of identifying whether the way we think the work is being done matches the way it is actually being done. This occurs at a number of levels and the primary driver is the creation and improvement of 'safe work', rather than the finding and correcting of 'unsafe work' and 'unsafe conditions'. One such process is called a PTO (Planned Task Observation).

The purpose of a Planned Task Observation is to identify and understand any differences between the way the work is intended to be done (as per the procedure/process) and the way it is actually done in the field so that we can learn and share when our people have identified better ways to do the work and, over time, align the Work-as-Done with the Work-As-Written through procedure improvement, sharing of learning, and behavioural changes if required.

The procedure that is to be the subject of the Planned Task Observation is reviewed prior to the observation. It is then taken into the field and a conversation started with workers, exploring how they are actually doing a task. Many topics may be covered during the conversation in order to get a sense of the level of understanding of the crew as to the control of risks and whether they align with the requirements of the procedure or not. Examples of the sorts of topic covered in PTOs could include:

- Assumptions by the workers about how the work is being conducted as compared to how the leader thinks it is in the procedure.
- Level of understanding of the responsibilities and activities of the team members for the task.
- What they think will go right? Or will go wrong?
- What they think about the procedures?

- Do they have plans for what to do if things go wrong?
- What they normally do when something unexpected happens?
- Has any drift occurred over time in the practice or the procedure?
- How stable their work normally is? Is it routine or does it require a high level of improvisation?
- What could happen that may influence their ability to follow the procedure?
- What they have to keep an eye on in the environment of the task? What they focus on?
- What level of Situational Awareness is needed, and on what, specifically?
- What is in the line of fire that can cause them injuries?
- What is going on in relation to critical controls that they should be considering, especially if there are some in the procedure of interest?
- If they had to describe what they are doing and how they are controlling risk, what they would say?
- Discussion about any anomalies in the workplace that do not appear to have been there before – anything that stands out as different.
- Do they have a plan if something does go wrong?
- What if …?
- How they built their mental model of the task prior to starting it (hazard, risk control, SIMOPs, etc.)?
- What must go right?
- What usually goes right here, what has gone well in the past?
- What will happen to them and the team if they do not control the risks?

The intent of these topics is to raise the level of conversation, through which it can be established whether the procedure has been followed, and more importantly why or why not. Of course, this is just a quick brain dump of engagement conversation topics possibilities. The intent is that these are not a set of questions but rather a conversation.

When we do a PTO and discover a part of a task that is being undertaken in a way that differs from what is outlined in the procedure or work instruction, we would conventionally do one of two things: one is to tell the worker to change their behaviour, and the other is to change the procedure to match the way the work is being done. We suggest that there is a third way that will better help to create safe work going forward. It is to undertake a micro-investigation – a mini learning study – what we call an LNW review based on what we have observed and only make changes after that if it makes sense and we understand the drivers for the way the work is being done.

Let's work through an example. You have decided to explore some scaffold construction work and have grabbed the scaffold procedure and had a look through it prior to going into the field. You have decided to focus on the controls scaffolders are using as they erect scaffold. You plan to look out for foot plates, harness use, attachment points, and how the scaffolders ensure no unauthorised people access the scaffolding during construction. You go into the field and start chatting with a work team building the scaffold. You notice that they have a piece of danger tape across the entrance but you recall that the procedure requires a drop bar to be installed rather than danger tape and so you start asking the scaffolders about this. They say that they quite often use danger tape as it is easier to duck under when they are going in and out of the scaffold as they are building it. Rather than making a big fuss about it and telling them to put a drop bar in, you decide to visit some other scaffolding erection activities going on in the plant to see what Work-As-Normal is – how others secure their scaffold entrances during construction. Over the next day or so, you visit five scaffold construction crews at work on the plant shutdown. Three out of the five had used danger tape, one had used a drop bar, and one simply had the scaffold tag removed from its holder as the control to prevent unwanted personnel from entering the scaffold.

The following morning you hold the scaffold crews back after the morning pre-shift meeting for 30 minutes and run an LNW review as a team activity. The intent is to explore and understand the variability amongst the crews as to how they secure scaffold entrances during construction. The format is a conversation with participation – hopefully – by all present to explore their level of understanding of what the normal practice is and more importantly why it is what it is. It turns out during the conversation that your business is the only one that they do work for that requires drop bars and all other companies only require danger tape or just leaving the scaffold tag out of its holder as a control to prevent unwanted people from accessing the scaffold as it is being built. They are simply not used to installing a drop bar – they forget to do it.

You then move the conversation into what they think can be done about it and what good practice would look like. The unanimous voice is a request for you to change the procedure as danger tape is in their opinion just as effective a control as a drop bar is. You understand their wish for simplicity and alignment with other companies and agree with their request. You walk away from the conversation, agreeing to change the procedure, and allow tape to be used in lieu of a drop bar. A half-hour conversation has resulted in a more accurate procedure, effective buy-in by those who control the risks (the scaffolders), and less of a gap between Work-As-Done and Work-As-Written. Everybody is happy.

Chapter 5

Assessing and Measuring Success

> Anything can be measured. If something can be observed in any way at all, it lends itself to some type of measurement method. No matter how 'fuzzy' the measurement is, it's still a measurement if it tells you more than you knew before.
>
> *How to Measure Anything*, **Douglas W. Hubbard**

I apologise to those of you who dislike audits. Much of this chapter may appear repetitive, mechanistic, and very audit-like. That this chapter has a look and a feel of an audit should not come as a surprise as audits are, by their very nature, tools of assessment and measurement. Also, the *Essentials of Safety* elements are in many ways similar to each other and so the assessing and measuring tools are also similar. There are principally two ways to view this chapter. You can use it as a formal set of measurements and audits to assess the health of the state of safety, or you can use it to trigger thoughts about how measurements could work for you and your business. Either way, I think the chapter contains enough guidance to whet your appetite for thinking about how you would tackle the assessing and measuring of the efficacy of the elements you are embedding. By measuring the *Essentials of Safety* elements, we aim to reduce the uncertainty around how embedded each is in the business. The first question we need to answer is whether it is legitimate to measure safety by measuring the *Essentials of Safety* elements. Measuring and assessing each of the elements in the ways

DOI: 10.1201/9781003181620-5

suggested here can, I believe, be viewed as a valid assessment of safety generally because we have established in previous chapters that if we get the 12 elements embedded across the business, we will be creating safe work through our people, through our leadership, by the systems we operate under, and as demonstrated by the workplace cultures we have. They are, after all, the essentials of safety, and measuring them is much better than lagging measurements such as injury rates and days since the last incident. As long as the measurement for each element reduces uncertainty with respect to that element, we should feel confident that we are heading in the right direction. The *Essentials of Safety* elements will have slightly different approaches to measurement and not all elements are measurable in a quantitative way. I think this is fine as many other aspects of business that are not safety related are assessed and measured in terms of qualitative measures in addition to quantitative ones.

The purpose of measurement is to gather information about the variable to be measured so that we reduce the uncertainty in the value of that variable. Generally, measuring is all about getting data in order to make decisions. Everything is measurable to this extent. The purpose of measuring the *Essentials of Safety* elements is to understand where they are up to and then to make decisions based on that understanding. Measurements will have errors – they all do. The question becomes one of whether the error in the measurement is sufficiently small so as to be swamped by the reduction in uncertainty the measurement has provided. In order to be able to measure a variable such as one of the *Essentials of Safety* elements, we need to explore a number of things for each element.

These include:

- What the observable variable or behaviours are.
- How to gather information on the observable variable.
- What the unexpected consequences of the measurement process may be.
- How we quantify the measurement.
- What the threshold value of the variable we consider to be an 'action level'.

It is usually important to go through a number of measurements before you start trying to deeply embed *Essentials of Safety* in order to establish a baseline. This way you can get an understanding as to whether you are witnessing improvement or not over subsequent measurements. I recommended that you run the measurements on all the elements for a few months prior to changing too much with respect to your approach to the *Essentials of Safety*.

For all of the *Essentials of Safety* elements, we are attempting to gather information through measurements in order to assist us in deciding whether the particular element is sufficiently embedded in the business so as to provide the right level of attention, leadership focus and behaviours, further implementation within systems, or discussion within the business.

In some cases, a detailed review of the data collected is made on a six-monthly basis. The data are summarised and the average 'score' computed. This will guide the senior leadership team in the business as to the direction of progress and highlight areas they wish to enhance and focus on. We will go through each of the elements, and then in the appendix to the chapter we provide some template worksheets to help you gather the information. These are also available on my website: www.raeda.com.au.

We must remember that like everything related to safety and especially to the essentials of safety, it is all about balance. To get a good sense of where you are, you do need to balance your assessing and measuring and not put all of your eggs in one basket. Each element is described but do try to keep it balanced (Figure 5.1).

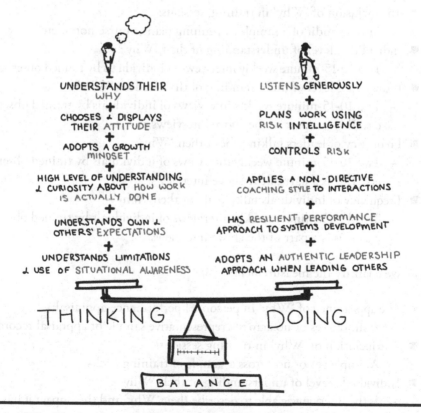

Figure 5.1   Keep your measurements balanced.

Thinking

Understands Their 'Why'

Observable variable/behaviour:

- The appearance of 'Why' in personnel performance appraisals.
- The inclusion of 'Why' in training sessions.
- Individual's level of understanding of their 'Why'.
- Individual's level of understanding of their leader's 'Why'.
- Frequency of leaders talking about their 'Why'.
- Frequency of individuals talking about their 'Why'.

Information-gathering process for the variable or behaviour:

- The appearance of 'Why' in personnel performance appraisals:
  - Annual audit of a sample of personnel performance appraisal records.
- The inclusion of 'Why' in training sessions:
  - Annual audit of a sample of training plans, course notes, etc.
- Individual's level of understanding of their 'Why':
  - Two 10-15-minute weekly interviews of individuals by trained observers.
- Individual's level of understanding of their leader's 'Why':
  - Two 10-15-minute weekly interviews of individuals by trained observers. This is a part of the above interviews.
- Frequency of leaders talking about their 'Why':
  - Two 10-15-minute weekly interviews of individuals by trained observers. This is a part of the above interviews.
- Frequency of individuals talking about their 'Why':
  - Two 10-15-minute weekly interviews of individuals by trained observers. This is a part of the above interviews.

How we quantify the measurement:

- The appearance of 'Why' in personnel performance appraisal:
  - A simple yes or no across a representative sample of appraisal records.
- The inclusion of 'Why' in training sessions:
  - A simple yes or no across a sample of training packs.
- Individual's level of understanding of their 'Why':
  - Is the interviewee able to describe their 'Why' and the impact it has on their work? (Score 1-5 as per below)

1. Participant does not understand what the question means.
2. Participant knows what is meant but is unable to describe their 'Why'.
3. Participant is able to describe their 'Why' but seems to need to think about it and/or make it up on the spot.
4. Participant is able to recall their 'Why' but not easily able to describe how it impacts their work.
5. Participant is able to recall their 'Why' and easily able to describe how it impacts how their work.

- Individual's level of understanding of their leader's 'Why':
  - Is the interviewee able to describe their leader's 'Why' and the impact it has on their work? (Score 1-5 as per below)
    1. Participant does not understand what the question means.
    2. Participant knows what is meant but is unable to describe their leader's 'Why'.
    3. Participant is able to describe their leader's 'Why' but seems to need to think about it and/or make it up on the spot.
    4. Participant is able to recall their leader's 'Why' but not easily able to describe how it impacts their work.
    5. Participant is able to recall their leader's 'Why' and easily able to describe how it impacts how their work.

- Frequency of leaders talking about their 'Why':
  - How often their leader, at any level, talk about their 'Why'. (Score 1-5 as per below)
    1. Never.
    2. Maybe once every few months.
    3. Probably once a month.
    4. Maybe once every couple of weeks.
    5. At least once per week.

- Frequency of individuals talking about their 'Why':
  - How often the interviewee talks about their 'Why'. (Score 1-5 as per below)
    1. Never.
    2. Maybe once every few months.
    3. Probably once a month.
    4. Maybe once every couple of weeks.
    5. At least once per week.

What is the threshold value of the variable we consider to be an 'action level':

For the two audit-related assessments, i.e. 'The appearance of "Why" in personnel performance appraisals' and 'The inclusion of "Why" in training sessions', it is suggested to have a sample size commensurate with the size of your business. An average of less than 75% 'Yes' would be a valid cause for consideration of action. As the remainder of the measurements are being undertaken during interview, a summary each six months will yield a sample size of between 50 and 60 scores. An action level for any of the measurements would be an average score over the six months of 3 or less. As a reminder, the measures undertaken during interviews are:

- Individual's level of understanding of their 'Why'.
- Individual's level of understanding of their leader's 'Why'.
- Frequency of leaders talking about their 'Why'.
- Frequency of individuals talking about their 'Why'.

Chooses and Displays Their Attitude

People's behaviour is usually consistent with, matches, or comes from their mindset and attitude. We need to establish whether people know what impact the way they behave and approach work can have on their well-being and that of others around them.

There are a number of ways to measure a person's attitude but they rely on being there at the time that the attitude is being displayed. The problem is that a person's attitude can be greatly affected by the act of measurement. Therefore, I feel the most effective way to measure 'Chooses and displays their attitude' is by using a Workplace Attitude (WPA) style modified guided interview process. Ray Valades, from Pepperdine University's Graduate School of Business, describes it well here (search for 'Ray Valades WPA' on www.gbr.pepperdine.edu).

The idea is to compare a person's *expectation* related to a topic to their *experience* of the topic and to express the result as a fraction. For example, if you talk about receiving feedback, the two statements scored could be:

- 'Ideally, I would expect to receive feedback on my attitude'.
- 'I actually receive feedback on my attitude'.

Topic	Expectation	Experience
To receive feedback on my attitude	'I agree that I would expect to receive feedback on my attitude' Score = 6	'I strongly disagree that I receive feedback on my attitude' Score = 1

The WPA score would be 1/6 = 0.17

The above scores are based on a conversation between the data collector and the interviewee. During the 10–15-minute conversation, the views of the interviewee are gathered on five topics. The interviewee's thoughts are assigned to one of the following levels:

1 Strongly disagree.
2 Slightly disagree.
3 Disagree.
4 Neutral.
5 Slightly agree.
6 Agree.
7 Strongly agree.

Topics discussed to determine the WPA score:

■ To receive feedback on my attitude.
■ To have time to think about my attitude before starting a task.
■ To be allocated tasks that match my capability and disposition.
■ My boss to shut down negative attitudes in others.
■ To have a work environment/team where those with negative attitudes have been weeded out.

The scores for both the *expectation* and the *experienced* are totalled for each interview and analysed over a six-month period. The WPA is the total *experienced* score divided by the total *expectation* score. For example:

	Expectation	Experienced
To receive feedback on my attitude	6	1
To have time to think about my attitude before starting a task	6	3
To be allocated tasks that match my capability and disposition	4	2
My boss to shut down negative attitudes in others	7	2
To have a work environment/team where those with negative attitudes have been weeded out	5	2
Total	28	10

WPA = 10/28 = 0.4

How to gather information on the observable variable:

■ Two data collectors undertake one interview per week each for 10-15 minutes and collate the data for the six-monthly review.

What the threshold value of the variable we consider to be an 'action level':

■ The action limit on the six-monthly review average WPA score is 0.60 and so if the average score is less than this, the element requires some consideration for action.

Adopts a Growth Mindset – including a Learning Mindset

Observable variable/behaviour:

■ Incident investigation/learning study reports that contain learning opportunities for the broader business.
■ Recruiting records that took growth mindset or fixed mindset into consideration during the hiring process.
■ Individual's performance appraisal records that include references to mindset in discussions and/or development plans.
■ People's level of agreement with:
  – The purpose of a workplace incident investigation is to learn.
  – If they want to, people can change their ability to learn new things.
  – People's level of intelligence can change over time and effort.

Information-gathering process for the observable variable or behaviour:

■ Incident investigation/learning study reports that contain learning opportunities for the broader business:
  – Annual audit of a sample of incident investigation/learning study reports.
■ Recruiting records that took growth mindset or fixed mindset into consideration during the hiring process:
  – Annual audit of a sample of recruitment records.
■ Individual's performance appraisal records that include references to mindset in discussions and/or development plans:
  – Annual audit of a sample of personnel performance appraisal records.
■ People's level of agreement with:
  – The purpose of a workplace incident investigation is to learn.

- If they want to, people can change their ability to learn new things.
- People's level of intelligence can change over time and effort.

Two 10-15-minute weekly interviews of individuals by trained observers.
How we quantify the measurement:

■ Incident investigation/learning study reports that contain learning opportunities for the broader business:
  - A simple yes or no across a sample of reports.
■ Recruiting records that took growth mindset, or fixed mindset, into consideration during the hiring process:
  - A simple yes or no across a sample of recruitment records.
■ Individual's performance appraisal records that include references to mindset in discussions and/or development plans:
  - A simple yes or no across a sample of an individual's performance appraisal records.
■ People's level of agreement with:
  - The purpose of a workplace incident investigation is to learn.
  - If they want to, people can change their ability to learn new things.
  - People's level of intelligence can change over time and effort.
    • Determine the level of agreement with each of the above questions using the scoring system:
      1 Strongly disagree.
      2 Slightly disagree.
      3 Disagree.
      4 Neutral.
      5 Slightly agree.
      6 Agree.
      7 Strongly agree.

What the threshold value of the variable we consider to be an 'action level':
The threshold value for the audits is 60%, and for the interview questions, based on a six-monthly review, it is an 'action level' of 4 or less.

Has a High Level of Understanding and Curiosity about How Work Is Actually Done

The measurements for this element are a mixture of annual audit and interviews. The interviews need to consider both worker- and leader-level interviewees.

Observable variable/behaviour:

- Incident investigation/learning study reports that contain meaningful inclusion of Work-As-Done, Work-As-Normal, and Work-As-Written.
- Level of understanding (leaders and workers) based on responses to the following questions:
  - What can you tell me about Work-As-Done, Work-As-Normal, and Work-As-Written?
  - Are Work-As-Done, Work-As-Normal, and Work-As-Written always the same or sometimes different?
  - If Work-As-Done, Work-As-Normal, and Work-As-Written are sometimes different, why?

Information-gathering process for the variable or behaviour:

- Incident investigation/learning study reports that contain meaningful inclusion of Work-As-Done, Work-As-Normal, and Work-As-Written:
  - Annual audit of a sample of incident investigation/learning study reports.
- Level of understanding (leaders and workers) based on responses to the above questions:
  - Two 10-15-minute weekly interviews of individuals by trained observers (one on a leader and one on a worker, alternating each week).

How we quantify the measurement:

- Incident investigation/learning study reports that contain meaningful inclusion of Work-As-Done, Work-As-Normal, and Work-As-Written:
  - A simple yes or no across a sample of reports.
- Level of understanding (leaders and workers) based on responses to the above questions:
  - With answers scored as a level of understanding: Scoring from 1 to 5 as follows:
    1 Very poor.
    2 Poor.
    3 Average.
    4 Good.
    5 Excellent.

What the threshold value of the variable we consider to be an 'action level':
- The threshold value for the audits is 60% and for the interview questions, based on a six-monthly review, it is an 'action level' of 3 or less.

Understands Their Own and Others' Expectations

In the case of expectations, we want to target leaders, their direct manager's, and their direct reports, and check for alignment in terms of expectations they have of themselves and of each other.

Observable variable/behaviour:

- Level of understanding of, and awareness of, expectations people have of themselves, expectations they have of others, and expectations others have of them.
- Alignment of these expectations within a three-level layer of the organisation.

Information-gathering process for the variable or behaviour:

- Week one: Interview a random individual.
- Week two: Interview the individual's direct manager.
- Week three: Interview one of the individual's direct report, if they have any.

How we quantify the measurement:

- Each week, ascertain whether the individuals are able to describe the expectation and through their descriptions of concrete examples, a score is given based on their level of understanding using the below scoring system:
  1. Very poor understanding.
  2. Poor understanding.
  3. Average understanding.
  4. Good understanding.
  5. Excellent understanding.
- At the end of the three-week cycle, score the overall alignment between the individual, their leader, and their direct report as follows:
  1. Very poor alignment.
  2. Poor alignment.
  3. Average alignment.

4. Good alignment.
5. Excellent alignment.

What the threshold values of the variable we consider to be an 'action level':

■ Six-monthly review of interview data:
  - Average of 3 or less requires consideration of action.
■ Six-monthly review of aggregated three-week cycle alignment data:
  - Average of 3 or less requires consideration of action.

Understands the Limitations and Use of Situational Awareness

Observable variable/behaviour:

■ Individuals' understanding of the limitations of situational awareness.
■ Individuals are observed approaching risk assessments during task-based risk assessments from various perspectives, including other people's perspectives.
■ Individuals observed talking about and sharing mental models of the work prior to starting it.
■ Individuals stopping and looking around when initially entering an area where work is planned to be done.
■ Individuals stopping and looking around in the middle of a task to see if anything has changed.
■ Individuals identifying the elements of the work or things within the workplace that they must pay attention to as they undertake a task.

Information-gathering process for the variable or behaviour:

■ Individuals' understanding of the limitations of situational awareness:
  - Two 10-15-minute weekly interviews of individuals by trained observers.
■ Individuals are observed approaching risk assessments during task-based risk assessments from various perspectives, including other people or other professions' perspectives.
■ Individuals observed talking about and sharing mental models of the work prior to starting it.
■ Individuals stopping and looking around when initially entering an area where work is planned to be done.

- Individuals stopping and looking around in the middle of a task to see if anything has changed.
- Individuals identifying the elements of the work or things within the workplace that they must pay attention to, as they undertake a task:
  - For all of the above, the two data collectors routinely visiting worksites, observing behaviour, and asking questions.

How we quantify the measurement:

- Individuals' understanding of the limitations of situational awareness:
  - Determine the level of understanding of what situational awareness is, as related to each of the three statements:
    - We cannot see everything.
    - We do not always see what is there (invisible Gorilla inattention blindness and inattention deafness).
    - Is all about knowing what is going on around us, and which bits are critical to keep an eye on.
  - Scoring from 1 to 5 as follows:
    1. Very poor understanding.
    2. Poor understanding.
    3. Average understanding.
    4. Good understanding.
    5. Excellent understanding.
- The remaining five sections related to the application of situational awareness in the field.
  - Scoring from 1 to 5 as follows:
    1. Very poor application of situational awareness.
    2. Poor application of situational awareness.
    3. Average application of situational awareness.
    4. Good application of situational awareness.
    5. Excellent application of situational awareness.

What the threshold value of the variable we consider to be an 'action level':

- Six-monthly review of interview data:
  - Average of 3 or less in either measure requires consideration of action.
- Six-monthly review of site field visit data:
  - Average of 3 or less in either measure requires consideration of action.

Doing

Listens Generously

Observable variable/behaviour, whilst listening:

- ■ Not interrupting:
  - − The listener is observed to wait a couple of seconds after the speaker stops talking.
  - − The listener maintains a passive or interested face whilst the speaker is talking.
- ■ Paying attention to the speaker, including not checking phone, email:
  - − The speaker looks at their phone during the conversation.
  - − The speaker does not react to beeps and rings on their phone.
  - − If the conversation is in an office, the speaker situates the conversation so that a screen does not form a barrier between the speaker and the listener, and does not look at it.
- ■ Using body language that tells the speaker they are listening:
  - − The speaker nods occasionally, maintains a reasonable level of eye contact, undertakes some mirroring of the speaker.
- ■ Giving time to the speaker:
  - − The speaker allows a reasonable time for silence after the speaker stops talking.
  - − The speaker asks questions such as 'What else?' 'What more is there?
- ■ Summarising before moving on and using the speaker's words in a response:
  - − Here is a conversation by way of example: A speaker says '… and so, the biggest issue is that the procedures that you want us to follow are written by people who have not got a clue about what the work is, how it needs to be done, what the important bits are, and sop the procedures cannot always be followed.' The listener might say: 'Okay, if I have this right, you're saying that the procedures are not easy to follow as they are written by people who don't know the work and that is why they are too hard to follow, right?' The speaker nods. The listener then goes on. 'Okay then, let's look at some specifics ….'.
- ■ Adding to the conversation:
  - − The listener starts their response by talking about the exact same aspect of the conversation that the speaker just espoused. For example, the listener starting a response to a speaker with 'and therefore …', 'and so …' or 'which leads onto…'

Information-gathering process for the variable or behaviour:

- Not interrupting.
- Paying attention to the speaker, including not checking phone, email.
- Using body language that tells the speaker they are listening.
- Giving time to the speaker.
- Summarising before moving on and using the speaker's words in a response.
- Adding to the conversation.

For all of the above behaviours, we undertake two separate activities. They are a combination of 'declared' and 'observed' assessments. The 'declared' are through a series of routine, random data-gathering conversations and the 'observed' assessments are through the installation of listening coaches in meetings. The trained data collectors undertake one random 15-20-minute interview conversation once each two weeks and sits in a meeting as a listening coach once per two weeks. The reason this element has longer interview times is that each measure has a different scale and we want to hear examples from the individual in each variable/behaviour. This takes time.

How we quantify the measurement:

For the 'declared' behaviour measurement (by two trained data collectors during interview conversations in the field – getting verbal thoughts from individuals as they relate observed behaviour of their leaders as the individual was talking with the leader):

Not interrupting. Scoring from 1 to 5 as follows:

1. The leader shows consistent interrupting behaviour.
2. There were quite a few examples of interruptions described and the individual was clearly impacted.
3. There were quite a few examples of interruptions described and the individual was not obviously impacted.
4. There were minimal (one or two) examples of interrupting and it did not appear to adversely impact the individual.
5. There was no interrupting at all described during the measurement.

Paying attention to the speaker, including not checking phone, email. Scoring from 1 to 5 as follows:

1. The leader shows consistent inattention to the individual during interactions.

2. There were quite a few examples of attention being diverted away from the individual and the individual was clearly impacted.
3. There were quite a few examples of inattention and the individual was not obviously impacted.
4. There were minimal (one or two) examples of a lack of attention by the leader and it did not adversely impact the individual.
5. The leader constantly pays attention to the individual.

Using body language that tells the speaker they are listening. Scoring from 1 to 5 as follows:

1. The leader's body language appears to the individual to be closed and disinterested.
2. There were quite a few examples of disinterested and closed body language and the individual was clearly impacted.
3. There were quite a few examples of disinterested and closed body language and the individual was not obviously impacted.
4. The leader's body language appeared reasonably engaging.
5. The leader's body language appeared engaging and the individual appeared to respond.

Giving time to the speaker. Scoring from 1 to 5 as follows:

1. The leader spoke over or immediately after the individual finished a sentence.
2. There were quite a few examples when the leader did not give a pause after the individual stopped talking and the speaker was clearly impacted.
3. There were quite a few examples when the leader did not give a pause after the individual stopped talking and it did not appear to adversely impact the individual.
4. There were minimal (one or two) examples of no, or minimum time given to the individual, and it did not appear to adversely impact the individual.
5. The leader always paused for at least a couple of seconds after the individual stopped talking before responding.

Summarising before moving on and using the speaker's words in a response. Scoring from 1 to 5 as follows:

1. The leader never summarised the last part of the individual's words before responding.

2. There were few examples of the leader summarising or using the individual's words during responses and interruptions and the individual was clearly impacted.
3. There were few examples of the leader summarising or using the individual's words during responses and interruptions and the individual was not obviously impacted.
4. There were quite a few examples of summarising by the leader and the listener using the individual's words during responses.
5. The leader nearly always summarised the last part of the individual's words before responding.

Adding to the conversation. Scoring from 1 to 5 as follows:

1. The leader's response bore no relationship to the last message of the individual.
2. The leader's response bore only a limited relationship to the last message of the individual.
3. The leader's response bore some relationship to the last message of the individual and the individual seemed to acknowledge this.
4. There were a number of examples of the leader attempting to link their response to the recent message of the individual.
5. There were numerous examples of the leader building on the words of the individual during the measurement.

For observed behaviour measurement (by a trained data collector acting as a listening coach and through observation in meetings – observing both speaker and listener)

Not interrupting. Scoring from 1 to 5 as follows:

1. The listener shows consistent interrupting behaviour during the measurement.
2. There were quite a few examples of interruptions and the speaker was clearly impacted.
3. There were quite a few examples of interruptions and the speaker was not obviously impacted.
4. There were minimal (one or two) examples of interrupting and it did not appear to adversely impact the speaker.
5. There was no interrupting at all during the measurement.

Paying attention to the speaker, including not checking phone, email. Scoring from 1 to 5 as follows:

1. The listener shows consistent inattention to the speaker during the measurement.
2. There were quite a few examples of attention being diverted away from the speaker, and the speaker was clearly impacted.
3. There were quite a few examples of inattention and the listener was not obviously impacted.
4. There were minimal (one or two) examples of a lack of attention by the listener and it did not appear to adversely impact the speaker.
5. The listener was constantly paying attention to the speaker during the measurement.

Using body language that tells the speaker they are listening. Scoring from 1 to 5 as follows:

1. The listener's body language appeared to be closed and disinterested.
2. There were quite a few examples of disinterested and closed body language and the speaker has clearly impacted.
3. There were quite a few examples of disinterested and closed body language and the listener was not obviously impacted.
4. The listener's body language appeared engaging, and the speaker did not appear to respond or acknowledge it.
5. The listener's body language appeared engaging and the speaker appeared to respond.

Giving time to the speaker. Scoring from 1 to 5 as follows:

1. The listener spoke over or immediately after the speaker finished a sentence.
2. There were quite a few examples when the listener did not give a pause after the speaker stopped talking and the speaker was clearly impacted.
3. There were quite a few examples when the listener did not give a pause after the speaker stopped talking and it did not appear to adversely impact the speaker.
4. There were minimal (one or two) examples of no, or minimum time given to the speaker and it did not appear to adversely impact the speaker.
5. The listener always paused for at least a couple of seconds after the speaker stopped talking before responding.

Summarising before moving on and using the speaker's words in a response. Scoring from 1 to 5 as follows:

1. The listener never summarised the last part of the speaker's words before responding.
2. There were few examples of the listener summarising or using the speaker's words during responses and interruptions and the speaker was clearly impacted.
3. There were few examples of the listener summarising or using the speaker's words during responses and interruptions and the speaker was not obviously impacted.
4. There were quite a few examples of summarising by the listener and the listener using the speaker's words during responses.
5. The listener nearly always summarised the last part of the speaker's words before responding.

Adding to the conversation. Scoring from 1 to 5 as follows:

1. The listener's response bore no relationship to the last message of the speaker.
2. The listener's response bore only a limited relationship to the last message of the speaker.
3. The listener's response bore some relationship to the last message of the speaker and the speaker seemed to acknowledge this.
4. There were a number of examples of the listener attempting to link their response to the recent message of the speaker.
5. There were numerous examples of the listener building on the words of the speaker during the measurement.

What the threshold value of the variable we consider to be an 'action level':

The threshold value of the variable we consider an 'action level' is 3 or less and requires action based on a six-monthly review of both activities.

Plans Work Using Risk Intelligence

'Plans' does not only apply to formal planning activities but also task-based planning on a day-to-day basis. This would be best seen during task preparation – during a task-based risk assessment creation activity, for example.

Observable variable/behaviour:

- Work teams discussing topics such as Efficiency-Thoroughness Trade-Off, Risk Intelligence, and a suitable wariness for the effectiveness of controls whilst planning a task that they are observed undertaking.
- The level of understanding of what Efficiency-Thoroughness Trade-Off, Risk Intelligence, and a suitable wariness for the effectiveness of controls means.

Information-gathering process for the variable or behaviour:

- Work teams discussing topics such as Efficiency-Thoroughness Trade-Off, Risk Intelligence, and a suitable wariness for the effectiveness of controls whilst planning a task that they are observed undertaking:
  - In the field discussions.
- The level of understanding of what Efficiency-Thoroughness Trade-Off, Risk Intelligence, and a suitable wariness for the effectiveness of controls means:
  - In the field discussions.

How we quantify the measurement:

- In both cases, trained data collectors visit individuals and teams in the work place and determine the level of practice and understanding concerning planning using the 1-5 scoring system below:
  1. Very poor understanding.
  2. Poor understanding.
  3. Average understanding.
  4. Good understanding.
  5. Excellent understanding.

As per many of the other *Essentials of Safety* elements, we use two trained data collectors who do all the interviews. They each do one random sample per week. It is important that the data are gathered from both formal planning activities and on-the-job task-to-task planning. We also undertake a six-monthly review of the results.

What the threshold value of the variable we consider to be an 'action level':
The threshold value of the variable we consider an 'action level' is as follows:
- Six-monthly review of interview data:
  - Average of 3 or less in either measure requires consideration of action.

Controls Risk

Observable variable/behaviour:

- ■ Trigger steps:
  - – Understanding of the term.
  - – Individuals and teams talking about trigger steps during task set up.
  - – Individuals stopping momentarily before any trigger steps.
- ■ Critical steps:
  - – Understanding of the term.
  - – Individuals and teams talking about critical steps during task set up.
  - – Individuals stopping momentarily before any critical steps.
- ■ Hazard identification:
  - – Understanding of the term.
  - – Individuals and teams talking about hazard identification during task set up.
- ■ Line-of-fire:
  - – Understanding of the term.
  - – Individuals and teams talking about line-of-fire hazards during task set up.
- ■ Simultaneous operations (SIMOPS):
  - – Understanding of the term.
  - – Individuals and teams talking about SIMOPS during task set up.
- ■ Escape plan/Preserving options:
  - – Understanding of the term.
  - – Individuals and teams talking about these during task set up.
  - – Individuals and teams talking about these during the task.
- ■ Telegraphing Deliberate Action:
  - – Understanding of the term.
  - – Individuals observed telegraphing deliberate actions.
- ■ Learning from Normal Work Reviews:
  - – Examples of completed LNW reviews:
    - • Observable in the sense that a report indicates an LNW review has been undertaken.
- ■ Procedural compliance/adaptation:
  - – Individuals and teams talking about how they are going to work compared to what is in the procedure during task set up, and including deviations in their task-based risk assessments.

Information-gathering process for the variable or behaviour:
■ Interviews of individuals by trained observers:
  – Two per week by trained observers undertaken at the work front looking for understanding and application.

How we quantify the measurement:

For each of the following, score from 1 to 5 using:
1. Very poor understanding and application.
2. Poor understanding and application.
3. Average understanding and application.
4. Good understanding and application.
5. Excellent understanding and application.
   • Trigger steps.
   • Critical steps.
   • Hazard ID.
   • Line-of-fire.
   • SIMOPS.
   • Escape plan/Preserving options.
   • Telegraphing Deliberate Action.
   • Procedural compliance/adaptation.
What the threshold value of the variable we consider to be an 'action level':
  – During the Six-monthly review of interview data, an average of 3 or less in either measure (understanding and application) requires consideration of action.

Applies a Non-Directive Coaching Style to Interactions

Observable variable/behaviour:

■ Using the GROW model in formal coaching sessions.
■ Using elements of coaching and the GROW model as a part of leadership behaviours.

Information-gathering process for the variable or behaviour:
The best measurement protocol is through establishing a formal coach-the-coach programme. As per previously described measurements, you will need two trained data collectors. In this case, they play the role of coach for the existing coaches and

leaders using a coaching style. These are long-term roles. These coaches' coaches focus not so much on direct one-on-one coaching with leaders but on helping the leader improve their coaching skills. Their focus is on using the GROW model with the coaches/leaders as they practice their coaching skills with individuals.

Each of these data collectors/coaches' coaches should undertake a coached coaching session once per week (with different leaders each week).

How we quantify the measurement:

Over a six-month period, the coaches' coaches keep an anonymous track of the quality of the leaders coaching skills by using scores from 1 to 5 as per below:

1. Very poor.
2. Poor.
3. Average.
4. Good.
5. Excellent.

What the threshold value of the variable we consider to be an 'action level'.

During a six-monthly review of the combined data collectors' data, a score of 3 or less requires consideration for action.

Has a Resilient Performance Approach to Systems Development

What the observable variable or behaviours are:

■ Evidence of resilience within the business's systems, covering:
  – The potential to respond.
  – The potential to monitor.
  – The potential to learn.
  – The potential to anticipate.

How to gather information on the observable variable.

Undertake annual audits of the business's systems assessing:

■ Assessing the potential to respond:
  – A list of everyday and unexpected events exists and are understood and updated.
  – A process that helps decide when an investigation is needed and to what level:
    • Both before and after a workplace safety incident.
  – A process that drives using a sound investigation process.

■ Assessing the potential to monitor:
   – A process that describes how to monitor for drift:
      • In its own elements.
      • In the implementation of those system elements.
   – A process that monitors the elements of the 'Safety Oscillation Model':
      • Impact of production or financial pressure on safety.
      • Impact of complacency on safety focus.
   – A process that monitors the interrelationships between the system elements and their impact on safety.
   – A process that described what is monitored in safety, and what safety-related Key Performance Indices (KPIs) exist.
   – Are KPIs:
      • Valid?
      • Reviewed?
      • Balanced between leading/lagging?
      • Resulting in changes?
■ Assessing the potential to learn:
   – A process that drives LNW reviews and/or 'what if' investigations to test the current and future systems:
      • A list of how many were done in the assessment period.
   – A description in the system that states how the business learns.
      • A process that describes how the business closes the loop from incident 'learning' outcome to actual learning and the testing of that learning across the business.
■ Assessing the potential to anticipate:
   – A process that drives regular consideration or forecasting of future changes such as in the legislative environment:
      • A process that drives the implementation of controls to address issues raised in the forecasting activities.
   – A process that describes how the business decides to change from a state of normal operation to a state of readiness when the conditions indicate that a crisis, disturbances, or failure is imminent.

How we quantify the measurement:

■ For the items described in each of the resilience measures above, score from 1 to 5 using:

1. Very poorly represented in the system/non-existent.
2. Poorly represented in the system/existent but not effective.

3. Average process but not well known.
4. Good process and reasonably well known.
5. Excellent process and implementation.

What is the threshold value of the variable we consider to be an 'action level'?

There is no specific threshold value for this measurement, although if you get a 3 or less in any of the measures, it is best if you consider action for that element in your annual planning processes.

Adopts an Authentic Leadership Approach to Leading Others

What the observable variable or behaviours are:

- Leaders are seen to behave consistently and according to their stated values and principles.
- Leaders show care in their behaviour and conversations with their team.
- Leaders' body language matches their words.
- Leaders routinely talk about their mistakes and the impact of them and what they have learned from them.
- Leaders do not stay aloof or in the knitting, but move between as needed.
- Leaders are observed to change approach when needed.
- Leaders talk about the fact that leadership starts with them.
- Leaders ask questions of others', and admit they are not the expert.

How to gather information on the observable variable.

The two main methods for carrying out the measurement are through directly scored observation of a leader interacting with someone who reports to them, and through interviews with employees and contractors across the business about their leaders. In line with many other elements you need to identify and train two data collectors for this element. They, on alternating weeks, either directly observe a randomly chosen leader in action or undertake a 10-15-minute interview with a randomly selected worker.

How we quantify the measurement:

For each of the behaviours described above, through direct observation of leaders and separately through interviews of workers, score each behaviour from 1 to 5:

1. Rarely.
2. Not often.
3. Sometimes.
4. Often.
5. Routinely.

What is the threshold value of the variable we consider to be an 'action level'?

Undertake a six-monthly review of the data from the two data collectors and have as an 'action level' an average of 3 or less in either measure.

In Summary

There could be unexpected consequences associated with undertaking the measurement and assessment processes as described above and so I will include a summary of them here.

A potentially negative consequence of the interview style assessment lies in people paying lip service to the question topic, supplying a canned answer with the interviewee giving the data collector the answers they think the data collectors want to hear. This can be minimised by ensuring a good explanation is given at the beginning of the interview and having trained data collectors. It can additionally be minimised by the data collector, asking clarification questions when an answer is given.

To minimise data collector bias, it is best to require data to be gathered by two trained data collectors. The idea here is to try to minimise individual bias in the measurement and in order to create consistency. For this reason, these data collectors should not be rotated out of their role very often.

They need to be trained to the level of knowing in sufficient detail 'Why' the element they are accountable to measure is important, what it is designed for, and how to undertake the measurement. Where the measurement is by way of interview, it needs to be undertaken as a conversation rather than a formal set of 'quoted' questions read off a list.

The interviewer must also abide by a few rules:

- No agreement or disagreement with the interviewee.
- No private opinions are to be discussed.
- No arguing with the interviewee.
- No suggestions are to be made to the interviewee, even for clarification.
- No leading questions or comments.

Most other unexpected consequences can be managed by building openness and trust in the process and in your people. This is of course the goal of the *Essentials of Safety* anyway so it is hoped that it will be self-reinforcing. As all of these measurements are designed to drive consideration of behaviour and action by the leaders in the business and are direct measures of safety in its creations, they should replace the existing lagging indicators currently used.

Chapter Appendix – Measurement Protocols Forms
Understands Their 'Why'

Does 'Why' appear in periodic performance appraisal write-up notes and development plans?

Sample No.	Yes	No	Comments
1			
2			
3			
4			
...			
18			
19			
20			
Total			

Annual review of a random sample of training packs for evidence of a discussion of the 'Why' of the training:

Sample No.	Yes	No	Comments
1			
2			
3			
4			
...			
18			
19			
20			
Total			

Understands their 'Why' interview worksheet – Circle appropriate response:

Topic	Score
If they are able to describe their 'Why' and the impact it has on their work	1. Participant does not understand what the question means 2. Participant knows what is meant but is unable to describe their 'Why' 3. Participant able to describe their 'Why' but seem to need to think about it and/or make it up on the spot 4. Participant able to recall their 'Why' but not able to easily describe how it impacts their work 5. Participant able to recall their 'Why' and easily able to describe how it impacts how they work
If they are able to describe their leader's 'Why' and the impact it has on their work	1. Participant does not understand what the question means 2. Participant knows what is meant but is unable to describe their 'Why' 3. Participant able to describe their 'Why' but seems to need to think about it and/or make it up on the spot 4. Participant able to recall their 'Why' but not easily describe easily how it impacts their work Participant able to recall their 'Why' and easily describe how it impacts how they work
How often their leaders (at any level) talk about their 'Why'	1. Never 2. Maybe once every few months 3. Probably once a month 4. Maybe once every couple of weeks 5. At least once a week or more
How often individuals talk about their 'Why'	1. Never 2. Maybe once every few months 3. Probably once a month 4. Maybe once every couple of weeks 5. At least once a week or more
Total	

Chooses and Displays Their Attitude

Circle appropriate response:

	Expectation	Experienced
To receive feedback on my attitude	1 – Strongly disagree 2 – Slightly disagree 3 – Disagree 4 – Neutral 5 – Slightly agree 6 – Agree 7 – Strongly agree	1 – Strongly disagree 2 – Slightly disagree 3 – Disagree 4 – Neutral 5 – Slightly agree 6 – Agree 7 – Strongly agree
To have time to think about my attitude before starting a task	1 – Strongly disagree 2 – Slightly disagree 3 – Disagree 4 – Neutral 5 – Slightly agree 6 – Agree 7 – Strongly agree	1 – Strongly disagree 2 – Slightly disagree 3 – Disagree 4 – Neutral 5 – Slightly agree 6 – Agree 7 – Strongly agree
To be allocated tasks that match my capability and disposition	1 – Strongly disagree 2 – Slightly disagree 3 – Disagree 4 – Neutral 5 – Slightly agree 6 – Agree 7 – Strongly agree	1 – Strongly disagree 2 – Slightly disagree 3 – Disagree 4 – Neutral 5 – Slightly agree 6 – Agree 7 – Strongly agree
My boss to shut down negative attitudes in others	1 – Strongly disagree 2 – Slightly disagree 3 – Disagree 4 – Neutral 5 – Slightly agree 6 – Agree 7 – Strongly agree	1 – Strongly disagree 2 – Slightly disagree 3 – Disagree 4 – Neutral 5 – Slightly agree 6 – Agree 7 – Strongly agree
To have a work environment/team where those with negative attitudes have been weeded out	1 – Strongly disagree 2 – Slightly disagree 3 – Disagree 4 – Neutral 5 – Slightly agree 6 – Agree 7 – Strongly agree	1 – Strongly disagree 2 – Slightly disagree 3 – Disagree 4 – Neutral 5 – Slightly agree 6 – Agree 7 – Strongly agree
Total		
WPA (Experienced total/Expectation total)		

Adopts a Growth Mindset

The number of incident investigation reports that have identified significant learning opportunities for the broader business:

Sample No.	Yes	No	Comments
1			
2			
3			
4			
5			
6			
7			
8			
9			
10			
Total			

A review of recent hires, where recruiters have identified those with growth mindsets and those with fixed mindsets and have used that knowledge meaningfully in their recruiting decisions:

Sample No.	Yes	No	Comments
1			
2			
3			
4			
5			
6			
7			
8			
9			
10			
Total			

A review of a sample of Performance Appraisal records and whether mindset has been discussed and actioned if needed in development plans:

Sample No.	Yes	No	Comments
1			
2			
3			
4			
5			
6			
7			
8			
9			
10			
Total			

Adopts a Growth Mindset – including a Learning Mindset

Circle appropriate response:

Topic	Score
The purpose of a workplace incident investigation is to learn	1 – Strongly disagree 2 – Slightly disagree 3 – Disagree 4 – Neutral 5 – Slightly agree 6 – Agree 7 – Strongly agree
If they want to, people can change their ability to learn new things	1 – Strongly disagree 2 – Slightly disagree 3 – Disagree 4 – Neutral 5 – Slightly agree 6 – Agree 7 – Strongly agree
People's level of intelligence can change over time and effort	1 – Strongly disagree 2 – Slightly disagree 3 – Disagree 4 – Neutral 5 – Slightly agree 6 – Agree 7 – Strongly agree
Total	

Has a High Level of Understanding and Curiosity about How Work Is Actually Done

Workplace incident investigation reports have meaningful inclusion of Work-As-Done, Work-As-Normal, and Work-As-Written?

Sample No.	Yes	No	Comments
1			
2			
3			
4			
5			
6			
7			
8			
9			
10			
Total			

Has a High Level of Understanding and Curiosity About How Work is Actually Done Interview Worksheet

Circle appropriate response:

Topic	Score
What can you tell me about Work-As-Done, Work-As-Normal, and Work-As-Written?	1 – Very poor understanding 2 - Poor understanding 3 - Average understanding 4 - Good understanding 5 – Excellent understanding
Are WAD, WAN, and WAW always the same, or sometimes different?	1 – Very poor understanding 2 – Poor understanding 3 – Average understanding 4 – Good understanding 5 – Excellent understanding
If WAD, WAN, and WAW are sometimes different, why?	1 – Very poor understanding 2 – Poor understanding 3 – Average understanding 4 – Good understanding 5 – Excellent understanding
Total	

Understands Their Own and Others' Expectations

Circle appropriate response:

Week 1: The individual. Look for people able to describe the expectation and giving concrete examples:

Topic	Score
Expectations they have of themselves	1 – Very poor understanding 2 – Poor understanding 3 – Average understanding 4 – Good understanding 5 – Excellent understanding
Expectations they have of others	1 – Very poor understanding 2 – Poor understanding 3 – Average understanding 4 – Good understanding 5 – Excellent understanding
Expectations others have of them	1 – Very poor understanding 2 – Poor understanding 3 – Average understanding 4 – Good understanding 5 – Excellent understanding
Total	

Week 2: The individual's direct manager. Look for people able to describe the expectation and giving concrete examples:

Topic	Score
Expectations they have of themselves	1 – Very poor understanding 2 – Poor understanding 3 – Average understanding 4 – Good understanding 5 – Excellent understanding
Expectations they have of others	1 – Very poor understanding 2 – Poor understanding 3 – Average understanding 4 – Good understanding 5 – Excellent understanding
Expectations others have of them	1 – Very poor understanding 2 – Poor understanding 3 – Average understanding 4 – Good understanding 5 – Excellent understanding
Total	

Week 3: The individual's direct report. Look for people able to describe the expectation and giving concrete examples:

Topic	Score
Expectations they have of themselves	1 – Very poor understanding 2 – Poor understanding 3 – Average understanding 4 – Good understanding 5 – Excellent understanding
Expectations they have of others	1 – Very poor understanding 2 – Poor understanding 3 – Average understanding 4 – Good understanding 5 – Excellent understanding
Expectations others have of them	1 – Very poor understanding 2 – Poor understanding 3 – Average understanding 4 – Good understanding 5 – Excellent understanding
Total	

Summary of the three-week cycle – circle appropriate level of alignment:

1. Very poor alignment.
2. Poor alignment.
3. Average alignment.
4. Good alignment.
5. Excellent alignment.

Understands the Limitations and Use of Situational Awareness

Circle appropriate response:

Topic	Score
We cannot see everything	1 – Very poor understanding 2 – Poor understanding 3 – Average understanding 4 – Good understanding 5 – Excellent understanding
We do not always see what is there (invisible Gorilla inattention blindness and inattention deafness)	1 – Very poor understanding 2 – Poor understanding 3 – Average understanding 4 – Good understanding 5 – Excellent understanding
Is all about knowing what is going on around us, and which bits are critical to keep an eye on	1 – Very poor understanding 2 – Poor understanding 3 – Average understanding 4 – Good understanding 5 – Excellent understanding
Total	

Understands the Limitations and Use of Situational Awareness Field Visit Worksheet

Circle appropriate response after observing the behaviour:

Topic	Score
Individuals are observed approaching risk assessments during Task-Based Risk Assessments from various perspectives, including other people or other professions' perspectives	1 – Very poor application of Situational Awareness 2 – Poor application of Situational Awareness 3 – Average application of Situational Awareness 4 – Good application of Situational Awareness 5 – Excellent application of Situational Awareness
Individuals observed talking about and sharing Mental Models of the work prior to starting it	1 – Very poor application of Situational Awareness 2 – Poor application of Situational Awareness 3 – Average application of Situational Awareness 4 – Good application of Situational Awareness 5 – Excellent application of Situational Awareness
Stopping and looking around when initially entering an area where work is planned to be done	1 – Very poor application of Situational Awareness 2 – Poor application of Situational Awareness 3 – Average application of Situational Awareness 4 – Good application of Situational Awareness 5 – Excellent application of Situational Awareness
Stopping and looking around in the middle of a task to see if anything has changed Identifying the elements of the work or things within the workplace that they must pay attention to, as they undertake a task	1 – Very poor application of Situational Awareness 2 – Poor application of Situational Awareness 3 – Average application of Situational Awareness 4 – Good application of Situational Awareness 5 – Excellent application of Situational Awareness
Total	

Listens Generously – For Observed Behaviour Measurement

Circle appropriate response:

Topic	Score
Not interrupting	1 – The listener shows consistent interrupting behaviour during the measurement 2 – There were quite a few examples of interruptions and the speaker was clearly impacted 3 – There were quite a few examples of interruptions and the speaker was not obviously impacted 4 – There were minimal (one or two) examples of interrupting and it did not appear to adversely impact the speaker 5 – There was no interrupting at all during the measurement
Paying attention to the speaker, including not checking phone, email	1 – The listener shows consistent inattention to the speaker during the measurement 2 – There were quite a few examples of attention being diverted away from the speaker and the speaker was clearly impacted 3 – There were quite a few examples of inattention and the listener was not obviously impacted 4 – There were minimal (one or two) examples of a lack of attention by the listener, and it did not appear to adversely impact the speaker 5 – The listener was constantly paying attention to the speaker during the measurement
Using body language that tells the speaker they are listening	1 – The listener's body language appeared to be closed and disinterested 2 – There were quite a few examples of disinterested and closed body language and the speaker was clearly impacted 3 – There were quite a few examples of disinterested and closed body language and the listener was not obviously impacted 4 – The listener's body language appeared engaging and the speaker did not appear to respond or acknowledge it 5 – The listener's body language appeared engaging and the speaker appeared to respond

Topic	Score
Giving time to the speaker	1 – The listener spoke over or immediately after the speaker finished a sentence 2 – There were quite a few examples when the listener did not give a pause after the speaker stopped talking and the speaker was clearly impacted 3 – There were quite a few examples when the listener did not give a pause after the speaker stopped talking and it did not appear to adversely impact the speaker 4 – There were minimal (one or two) examples of no, or minimum time given to the speaker and it did not appear to adversely impact the speaker 5 – The listener always paused for at least a couple of seconds after the speaker stopped talking before responding
Summarising before moving on and using the speaker's words in a response	1 – The listener never summarised the last part of the speaker's words before responding 2 – There were few examples of the listener summarising or using the speaker's words during responses and interruptions and the speaker was clearly impacted 3 – There were few examples of the listener summarising or using the speaker's words during responses and interruptions, and the speaker was not obviously impacted 4 – There were quite a few examples of summarising by the listener and the listener using the speaker's words during responses 5 – The listener nearly always summarised the last part of the speaker's words before responding
Adding to the conversation	1 – The listener's response bore no relationship to the last message of the speaker 2 – The listener's response bore only a limited relationship to the last message of the speaker 3 – The listener's response bore some relationship to the last message of the speaker and the speaker seemed to acknowledge this 4 – There was a number of examples of the listener attempting to link their response to the recent message of the speaker 5 – There were numerous examples of the listener building on the word of the speaker during the measurement
Total	

Listens Generously – For Declared Behaviour Measurement

Circle appropriate response:

Topic	Score
Not interrupting	1 – The leader shows consistent interrupting behaviour 2 – There were quite a few examples of interruptions described and the individual was clearly impacted 3 – There were quite a few examples of interruptions described and the individual was not obviously impacted 4 – There were minimal (one or two) examples of interrupting, and it did not appear to adversely impact the individual 5 – There was no interrupting at all described during the measurement
Paying attention to the speaker, including not checking phone, email	1 – The leader shows consistent inattention to the individual during interactions 2 – There were quite a few examples of attention being diverted away from the individual and the individual was clearly impacted 3 – There were quite a few examples of inattention and the individual was not obviously impacted 4 – There were minimal (one or two) examples of a lack of attention by the leader, and it did not adversely impact the individual 5 – The leader constantly pays attention to the individual
Using body language that tells the speaker they are listening	1 – The leader's body language appears to the individual to be closed and disinterested 2 – There were quite a few examples of disinterested and closed body language and the individual was clearly impacted 3 – There were quite a few examples of disinterested and closed body language, and the individual was not obviously impacted 4 – The leader's body language appeared reasonably engaging 5 – The leader's body language appeared engaging and the individual appeared to respond

Topic	Score
Giving time to the speaker	1 – The leader spoke over or immediately after the individual finished a sentence 2 – There were quite a few examples when the leader did not give a pause after the individual stopped talking and the speaker was clearly impacted 3 – There were quite a few examples when the leader did not give a pause after the individual stopped talking and it did not appear to adversely impact the individual 4 – There were minimal (one or two) examples of no, or minimum time given to the individual and it did not appear to adversely impact the individual 5 – The leader always paused for at least a couple of seconds after the individual stopped talking before responding
Summarising before moving on and using the speaker's words in a response	1 – The leader never summarised the last part of the individual's words before responding 2 – There were few examples of the leader summarising or using the individual's words during responses and interruptions and the individual was clearly impacted 3 – There were few examples of the leader summarising or using the individual's words during responses and interruptions and the individual was not obviously impacted 4 – There were quite a few examples of summarising by the leader and the listener using the individual's words during responses. 5 – The leader nearly always summarised the last part of the individual's words before responding
Adding to the conver-sation	1 – The leader's response bore no relationship to the last message of the individual 2 – The leader's response bore only a limited relationship to the last message of the individual 3 – The leader's response bore some relationship to the last message of the individual and the individual seemed to acknowledge this 4 – There were a number of examples of the leader attempting to link their response to the recent message of the individual 5 – There were numerous examples of the leader building on the word of the individual during the measurement
Total	

Plans Work Using Risk Intelligence

Circle appropriate response:

Topic	Score
How they planned the task they are currently doing - looking for discussion on: ETTO, Risk Intelligence, a suitable wariness for the effectiveness of controls	1 – Very poor understanding 2 – Poor understanding 3 – Average understanding 4 – Good understanding 5 – Excellent understanding
Their level of understanding of what Efficiency-Thoroughness Trade-Off (ETTO), Risk Intelligence, and a suitable wariness for the effectiveness of controls means	1 – Very poor understanding 2 – Poor understanding 3 – Average understanding 4 – Good understanding 5 – Excellent understanding
Total	

Controls Risk

Circle appropriate response:

Topic discussed within the conversation	Score
Trigger steps	1 – Very poor understanding and application 2 – Poor understanding and application 3 – Average understanding and application 4 – Good understanding and application 5 – Excellent understanding and application
Critical steps	1 – Very poor understanding and application 2 – Poor understanding and application 3 – Average understanding and application 4 – Good understanding and application 5 – Excellent understanding and application
Hazard ID	1 – Very poor understanding and application 2 – Poor understanding and application 3 – Average understanding and application 4 – Good understanding and application 5 – Excellent understanding and application
Line-of-fire	1 – Very poor understanding and application 2 – Poor understanding and application 3 – Average understanding and application 4 – Good understanding and application 5 – Excellent understanding and application
SIMOPS	1 – Very poor understanding and application 2 – Poor understanding and application 3 – Average understanding and application 4 – Good understanding and application 5 – Excellent understanding and application
Escape plan/Preserving options	1 – Very poor understanding and application 2 – Poor understanding and application 3 – Average understanding and application 4 – Good understanding and application 5 – Excellent understanding and application

Topic discussed within the conversation	Score
Telegraphing deliberate action	1 – Very poor understanding and application 2 – Poor understanding and application 3 – Average understanding and application 4 – Good understanding and application 5 – Excellent understanding and application
Procedural compliance/ adaptation	1 – Very poor understanding and application 2 – Poor understanding and application 3 – Average understanding and application 4 – Good understanding and application 5 – Excellent understanding and application
Total	

Applies a Non-Directive Coaching Style to Interactions

Circle appropriate response and comment

GROW model element	Score	Comment
Goal	1 – Very poor 2 – Poor 3 – Average 4 – Good 5 – Excellent	
Reality	1 – Very poor 2 – Poor 3 – Average 4 – Good 5 – Excellent	
Options	1 – Very poor 2 – Poor 3 – Average 4 – Good 5 – Excellent	
Wrap up	1 – Very poor 2 – Poor 3 – Average 4 – Good 5 – Excellent	
Total		

Has a Resilient Performance Approach to Systems Development

Circle appropriate response:

Topic	
Assessing the potential to respond:	
A list of everyday and unexpected events exists and are understood and updated	1 – Very poorly represented in the system/non-existent 2 – Poorly represented in the system/existent but not effective 3 – Average process but not well known 4 – Good process and reasonably well known 5 – Excellent process and implementation
A process that helps decide when an investigation is needed and to what level - both before and after a workplace safety incident	1 – Very poorly represented in the system/non-existent 2 – Poorly represented in the system/existent but not effective 3 – Average process but not well known 4 – Good process and reasonably well known 5 – Excellent process and implementation
A process that drives using a sound investigation process	1 – Very poorly represented in the system/non-existent 2 – Poorly represented in the system/existent but not effective 3 – Average process but not well known 4 – Good process and reasonably well known 5 – Excellent process and implementation
Assessing the potential to monitor:	
A process that describes how to monitor for drift	1 – Very poorly represented in the system/non-existent 2 – Poorly represented in the system/existent but not effective 3 – Average process but not well known 4 – Good process and reasonably well known 5 – Excellent process and implementation

Topic	
A process that monitors the elements of the 'Safety Oscillation Model' • Impact of production or financial pressure on safety • Impact of complacency on safety focus	1 – Very poorly represented in the system/ non-existent 2 – Poorly represented in the system/existent but not effective 3 – Average process but not well known 4 – Good process and reasonably well known 5 – Excellent process and implementation
A process that monitors the interrelationships between the system elements and their impact on safety	1 – Very poorly represented in the system/ non-existent 2 – Poorly represented in the system/existent but not effective 3 – Average process but not well known 4 – Good process and reasonably well known 5 – Excellent process and implementation
A process that described what is monitored in safety, and what safety-related KPIs exist	1 – Very poorly represented in the system/ non-existent 2 – Poorly represented in the system/existent but not effective 3 – Average process but not well known 4 – Good process and reasonably well known 5 – Excellent process and implementation
Are KPIs valid, reviewed, balanced between leading/ lagging, result in changes?	1 – Very poorly represented in the system/ non-existent 2 – Poorly represented in the system/existent but not effective 3 – Average process but not well known 4 – Good process and reasonably well known 5 – Excellent process and implementation
Assessing the potential to learn:	
A process that drives LNW reviews and/or 'what if', investigations to test the current and future systems, including how many were done	1 – Very poorly represented in the system/ non-existent 2 – Poorly represented in the system/existent but not effective 3 – Average process but not well known 4 – Good process and reasonably well known 5 – Excellent process and implementation

Topic	
A description in the system that states how the business learns	1 – Very poorly represented in the system/ non-existent 2 – Poorly represented in the system/existent but not effective 3 – Average process but not well known 4 – Good process and reasonably well known 5 – Excellent process and implementation
A process that describes how the business closes the loop from incident 'learning' out come to actual learning and the testing of that learning across the business	1 – Very poorly represented in the system/ non-existent 2 – Poorly represented in the system/existent but not effective 3 – Average process but not well known 4 – Good process and reasonably well known 5 – Excellent process and implementation
Assessing the potential to anticipate:	
A process that drives regular consideration or forecasting of future changes such as in the legislative environment	1 – Very poorly represented in the system/ non-existent 2 – Poorly represented in the system/existent but not effective 3 – Average process but not well known 4 – Good process and reasonably well known 5 – Excellent process and implementation
A process that drives the implementation of controls to address issues raised in the forecasting activities	1 – Very poorly represented in the system/ non-existent 2 – Poorly represented in the system/existent but not effective 3 – Average process but not well known 4 – Good process and reasonably well known 5 – Excellent process and implementation
A process that describes how the business decides to change from a state of normal operation to a state of readiness when the conditions indicate that a crisis, disturbance, or failure is imminent	1 – Very poorly represented in the system/ non-existent 2 – Poorly represented in the system/existent but not effective 3 – Average process but not well known 4 – Good process and reasonably well known 5 – Excellent process and implementation
Total	

Adopts an Authentic Leadership Approach When Leading Others

Circle appropriate response – through direct scored observation of a leader interacting with someone who reports to them:

Topic 'That leaders ...'	Score
Are seen to behave consistently and according to their stated values and principles	1 – Rarely 2 – Not often 3 – Sometimes 4 – Often 5 – Routinely
Show care in their behaviour and conversations with their team	1 – Rarely 2 – Not often 3 – Sometimes 4 – Often 5 – Routinely
Their body language matches their words	1 – Rarely 2 – Not often 3 – Sometimes 4 – Often 5 – Routinely
Routinely talk about their mistakes and the impact of them and what they have learned from them	1 – Rarely 2 – Not often 3 – Sometimes 4 – Often 5 – Routinely
Do not stay aloof or in the knitting, but move between as needed	1 – Rarely 2 – Not often 3 – Sometimes 4 – Often 5 – Routinely
Are observed to change approach when needed	1 – Rarely 2 – Not often 3 – Sometimes 4 – Often 5 – Routinely

Topic 'That leaders ...'	Score
Talk about the fact that leadership starts with them	1 – Rarely 2 – Not often 3 – Sometimes 4 – Often 5 – Routinely
Ask questions of others and admit they are not the expert	1 – Rarely 2 – Not often 3 – Sometimes 4 – Often 5 – Routinely
Total	

Adopts an Authentic Leadership Approach When Leading Others' Interview Worksheet

Circle appropriate response - through interviews with employees and contractors across the business about their leaders:

Topic 'That leaders ...'	Score
Are seen to behave consistently and according to their stated values and principles	1 – Rarely 2 – Not often 3 – Sometimes 4 – Often 5 – Routinely
Show care in their behaviour and conversations with their team	1 – Rarely 2 – Not often 3 – Sometimes 4 – Often 5 – Routinely
Their body language matches their words	1 – Rarely 2 – Not often 3 – Sometimes 4 – Often 5 – Routinely
Routinely talk about their mistakes and the impact of them and what they have learned from them	1 – Rarely 2 – Not often 3 – Sometimes 4 – Often 5 – Routinely
Do not stay aloof or in the knitting, but move between as needed	1 – Rarely 2 – Not often 3 – Sometimes 4 – Often 5 – Routinely
Are observed to change approach when needed	1 – Rarely 2 – Not often 3 – Sometimes 4 – Often 5 – Routinely

Topic 'That leaders ...'	Score
Talk about the fact that leadership starts with them	1 – Rarely 2 – Not often 3 – Sometimes 4 – Often 5 – Routinely
Ask questions of others and admit they are not the expert	1 – Rarely 2 – Not often 3 – Sometimes 4 – Often 5 – Routinely
Total	

Conclusion

This book is not about safety. This book is all about people. I came to this via the following path: we know that safety has a strong link with leadership. We also know that leadership is about relationship. And of course, relationship is all about people. To add to this, and to further the interrelationships between individuals, leaders, and leadership, the systems we use at work, and the workplace cultures, we need to recognise that people are individuals, people are leaders, people create the systems within which we operate, and people through their shared basic assumptions, beliefs, values, ideals, and observed behaviours create the workplace cultures. This all combined together to tell me that safety is all about people. This book is all about people – leaders at all levels – and the things for them to think about and things for them to do that will result in creating safety. This is safety, not just in the sense of outcomes of work being a lack of harm but the in-built, Created Safety that is all about how we work and how we achieve our goals.

In this work, I have tried to explore the idea that our leadership, regardless of our position in our business, or university, or hospital, or wherever we call home for our work, and regardless of what sector our business or undertaking is in, can unlock the common drivers that underpin the work of the *Individual*, *Leaders* and *Leadership*, the *Systems* we use, and the workplace *Cultures* in which we work. Through how we think and what we do, we can derive workplace satisfaction and success by thinking about why we do what we do, approaching work with a suitable attitude and with a mindset that is all about growth and learning, have a good understanding of how work is actually done, have a sound understanding of, and sharing of our own and others expectations, and having a shared mental model of work. This is all balanced with listening generously, planning well, controlling risk, using a coaching style in our leadership, developing systems thoughtfully, and being an authentic leader. Doing all of this will result in changes to the way you think about your job and how you do your job.

DOI: 10.1201/9781003181620-102

I am confident that you will not agree with everything I have written here. That is as it should be. I absolutely do not expect you to be of the same mind as me. What you have read are my views and thoughts based on my own 43 odd years of work in a large number of industries mixed with, and greatly influenced by many, if not all, of those mentioned in the bibliography.

I am not a supporter of blindly following a recipe. I know that I called this book a recipe book earlier; however, I tried to use that word with caution. As I said at the start, this is a book designed to prompt you to think, to think about what I have said, to think about what each and every element means to you and your industry, to think about what they might look like if they were embedded in your workplace and in your leadership style, and then to think about how you might apply them to a greater or lesser degree in the way you balance your thinking and doing in the creation of safe work.

If there is one take away that I want you to walk away from this book with it is this: safety is not about the outcome. Safety is created each and every time we think about work and every time we do work.

To add action to the words of Andrew Roberts from the quote at the start of the Introduction many pages in the past: *'If you have an important point to make, don't try to be subtle or clever. Use a pile driver. Hit the point once. Then come back and hit it again. Then hit it a third time.'* I will hit you with my message for the final time.

The *Essentials of Safety* are a set of elements that permeate through the workforce at all levels. They talk to the aspects of the Individual, Leaders and Leadership, the Systems we use, and the workplace Cultures. It is typified by a state of balance, where, driven through strong relationships and trust, everybody maintains a balance between thinking and doing:

Thinking
 – Understands Their 'Why'.
   • The more we understand our 'Why', the more we are able to be ourselves. The more we are able to be ourselves, the more effective we become. Our 'Why' is our underlying purpose. It is also the underlying purpose of what we do and how we do it.
 – Chooses and displays their attitude.
   • We do not always get to choose what we do, but we always have a choice as to how we view what we do – how we react.
 – Adopts a growth mindset – including a learning mindset.
   • People with a growth mindset understand that they can learn, change, adapt, and improve. People with a fixed mindset are the opposite. My advice is to go with the growth mindset.

- Has a high level of understanding and curiosity about how work is actually done.
  - We all need to understand that the way we think work is done out there is not always the same as the way it is actually done.
- Understands their own and others' expectations.
  - We all have expectations of ourselves, and of those around us. It is important to understand these expectations and to share them.
- Understands the limitations and use of Situational Awareness.
  - We are simply not able to keep an eye on absolutely everything around us. We are very selective observers. Keeping shared mental models across a team is a great way of jointly keeping an eye on what is important.

Doing
- Listens generously.
  - Listening is the most important skill a leader can possess. Paying attention to, being curious about, and authentically and actively listening are critical activities for any leader.
- Plans work using Risk Intelligence.
  - Planning is such a foundation activity. We need to get it right at many levels. Understanding the trade-offs between efficiency and thoroughness, understanding and applying Risk Intelligence, and having a suitable wariness for the effectiveness of controls can all help us keep our planning more effective.
- Controls risk.
  - Preserving options, understanding and managing critical and trigger steps, telegraphing deliberate action, being mindful, and ensuring the right tools and equipment are used, all help controlling risk.
- Applies a non-directive coaching style to interactions.
  - Using the GROW model, applying a non-directive coaching style, asking more than telling, learning, and practising coaching helps everyone be the best they can be.
- Has a Resilient Performance approach to systems development.
  - Build systems that help us make sure things go right, tell us what to look for, and monitor so that things are going to stay going right, learn from the past.
- Adopts an authentic leadership approach when leading others.

- This is really the guts of *Essentials of Safety*. Truly and authentically caring, setting context and framework, being ourselves only more so, intent-based and authentic leadership is what it is all about.

I fully acknowledge that becoming great at safety, if such a thing can truly be considered or said, is not ever as simple as reading a book such as *Essentials of Safety* and then simply applying it. In my view that would clearly not be the best thing to do. My recommendation is for you to think about what you have read, explore your own ideas and thoughts, your own business and industry, and then work out how it could look for you. You should, using the chapter on barriers and their remedies, understand what may present to you in your situation that could appear to prevent you from applying a particular element. You can then work through the remedies I have supplied, or better yet, use them as thinking fodder to come up with your own remedies to your barriers.

It is important to remember here that ideally all the 12 elements should be applied all of the time. But in practice some may present more as conversations than as actual practices, at least initially. Pick the elements that you like and that align better to your 'Why' to start with and embed them, expanding to others as you go. You can pick and choose the leadership practices and routines from Chapter 2 that you like. The options for activities were designed to be numerous and again designed to provide sufficient ideas so that you can explore them and then think about how and what you will do to further the element in your own leadership style, in addition to promoting ideas and thoughts in others' leadership behaviours. I fully acknowledge that the assessing and measuring success chapter will be viewed by some as a bit over the top. It is designed to be used as is for a full analysis of the situation but can also be used in a more minimalist way by using some of the assessing and measuring ideas and adapting them to your needs.

At the end of the day, I wanted this book to be in service of, and in support of, my 'Why'. I mentioned it earlier: 'To share ideas, concepts and practicalities in safety and leadership with as many people as will listen, so that people start to think differently and positively about the why, what and how of the things they do in both leadership and safety.'

I sincerely hope I have achieved this and welcome your thoughts and conversations on anything that you feel relates to this work at Ian.long@raeda.com.au

The worksheets in the appendix to Chapter 5 are also available on my website: www.raeda.com.au

Must-Read Bibliography: Books and Articles

1. *A Life in Error: From Little Slips to Big Disasters.* James Reason. Ashgate 2013.
2. *Accident Analysis and Hazard Analysis for Human and Organizational Factors.* Margaret V. Stringfellow. Partial fulfillment of PhD in Aeronautics and Astronautics. Internet.
3. *After Harm; Medical Error and the Ethics of Forgiveness.* Berlinger, N. Baltimore, MD: Johns Hopkins University Press 2005.
4. *An Introduction to the 5 Phases of HOP Integration.* Andrea Baker. www .SafetyDifferently.com.
5. *Behind Human Error.* 2nd ed. Sidney Dekker, together with David D. Woods, Richard Cook, Leile Johannsen and Nadne Sarter. Ashgate 2010.
6. *Better Questions: An Applied Approach to Operational Learning.* Todd Conklin. CRC Press 2016.
7. *Bury My Heart at Conference Room B: The Unbeatable Impact of Truly Committed Managers.* Stan Slap. Portfolio Penguin 2010.
8. *Close Calls: Managing Risk and Resilience in Airline Flight Safety.* Carl Mccrae. Palgrave Macmillan 2014.
9. *Coaching for Performance: Growing Human Potential and Purpose – The Principles and Practice of Coaching and Leadership.* John Whitmore. Nichols Brealey Publishing 1992.
10. *Cognitive Issues Associated with Process Safety and Environmental Incidents.* Report Number 460. International Association of Oil and Gas Producers. July 2012.
11. *Columbia Accident Investigation Board Report Volume 1.* August 2003. NASA.
12. Report Volume 1; 2003, NASA and the Government Printing Office of the USA (Obtained from www.nasa.gov/columbia/home/CAIB_Vol1.html).
13. *Comm Check: The Final Flight of Shuttle Columbia.* Michael Cabbage and William Harwood. Simon and Schuster 2008.
14. *Command and Control: Nuclear Weapons, the Damascus Accident, and the Illusion of Control.* Eric Schlosser. Penguin 2014.
15. *Controlling Risk: Thirty Techniques for Operating Excellence.* Jim Wetherbee. Morgan James Publishing 2019.
16. *Crucial Conversations: Tools for Talking When the Stakes are High.* Kerry Patterson, Joseph Grenny, Ron McMillan, and Al Switzler. McGraw-Hill 2009.
17. *Deep Leadership: Essential Insights from High-Risk Environments.* Joe MacInnes. Alfred A. Knopf Canada Publishing 2012.

18. *Designing to Avoid Disaster: The Nature of Fracture-Critical Design.* Thomas Fisher. Routledge 2012.
19. *Disastrous Decisions: The Human Organisational Causes of the Gulf of Mexico Blowout.* Andrew Hopkins.
20. *Discover Your True North: Becoming an Authentic Leader.* Bill George. Wiley 2015.
21. *Dispelling the Myths and Rediscovering the Lost Art of Listening.* Richard Mullender. Griffen Professional Business & Training Services 2012.
22. *Drift into Failure: From Hunting Broken Components to Understanding Complex Systems.* Sidney Dekker. Ashgate 2011.
23. DuPont 1914 Handout to employees.
24. *Effective Coaching: Lessons from the Coach's Coach.* 3rd ed. Myles Downey. Cengage 2003.
25. *Engineering a Safer World: Systems Thinking Applied to Safety (Engineering Systems).* Nancy G. Leveson. MIT Press 2016.
26. *Everything is Obvious Once You Know the Answer: How Common Sense Fails Us.* Duncan Watts. Atlantic Books 2011.
27. *Failure to Learn: The BP Texas City Refinery Disaster.* Andrew Hopkins. CCH Australia 2015.
28. *Feynman's Appendix to the Rogers Commission Report on the Space Shuttle Challenger Accident: Personal Observation on the Reliability of the Shuttle.* R.P. Feynman. Internet.
29. *Find Your Why: A Practical Guide for Discovering Purpose for You and Your Team.* Simon Sinek. Penguin Random House 2017.
30. *Fish.* Stephen C. Lundin, Harry Paul and John Christensen. Hodder 2000.
31. *Foundations of Safety Science: A Century of Understanding Accidents and Disasters.* CRC Press. Sidney Dekker.
32. *FRAM: Functional Resonance Analysis Method: Modelling Complex Socio-Technical Systems.* Erik Hollnagel. Ashgate 2012.
33. *Friendly Fire: The Accidental Shootdown of US Black Hawks over Northern Iraq.* Snook, S.A. Princeton University Press 2000.
34. *Good Leaders Ask Great Questions: Your Foundation for Successful Leadership.* John Maxwell. Center Street 2014.
35. *How Long Throughout the Millennium Remain in the Collective Memory.* Vaclav Fanta, Miroslav Salek, and Petr Sklenicka. Nature Communications 10, 1105 (2019).
36. *How to Measure Anything: Finding the Value of Intangibles in Business.* 3rd Ed. Douglas W. Hubbard. Wiley 2014.
37. *Human Error.* James Reason. North-Holland 1990.
38. *Human Factors and Ergonomics in Practice: Improving System Performance and Human Well-Being in the Real World.* Ed. Steven Shorrock and Claire Williams. CRC Press 2017.
39. *Humble Enquiry: The Gentle Art of Asking Instead of Telling.* Edgar Schein. Berrett-Koehler 2013.
40. *In Pursuit of Foresight: Disaster Incubation Theory Re-imagined.* Mike Lauder. Routledge 2016.
41. *Investigation of the Challenger Accident: Report of the Committee on Science and Technology House of Representatives.* US Government Printing Office 1986.
42. *It's Not About the Shark: How to Solve Unsolvable Problems.* David Niven. Icon Books 2015.

43. *Investigative Interviewing: Psychology and Practice.* Rebecca Milne and Ray Bull. John Wiley and Sons 1999.
44. *Just Culture: Balancing Safety and Accountability.* Sidney Dekker. Ashgate 2012.
45. *Leaders Eat Last: Why Some Teams Pull Together and Others Don't.* Simon Sinek. Portfolio Print 2014.
46. *Leadership Can Be Taught: A Bold Approach for a Complex World.* Sharon Daloz Parks. Harvard Business School Press 2005.
47. *Leadership is Language: The Hidden Power of What You Say – and What You Don't.* L. David Marquet. Penguin Books 2020.
48. *Learning from High Reliability Organisations.* Andrew Hopkins. CCH Australia 2014.
49. *Lessons from Gretley: Mindful Leadership and the Law.* Andrew Hopkins. CCH Australia 2007.
50. *Lessons from Longford: The Esso Gas Plant Explosion.* Andrew Hopkins. CCH Australia 2000.
51. *Man-Made Disasters. Barry Turner and Nick Pidgeon.* Butterworth-Heinemann 1997.
52. *Managing the Risk of Organizational Accidents.* James Reason. Ashgate 1997.
53. *Managing the Unexpected: Sustained Performance in a Complex World.* 3rd Ed. Karl Weick and Kathleen Sutcliffe. Josey-Bass 2015.
54. *Mastering Coaching: Practical Insights for Developing High Performance.* Max Landsberg. Profile Books 2015.
55. *Midnight in Chernobyl: The Untold Story of the World's Greatest Nuclear Disaster.* Adam Higginbotham. Random Press 2019.
56. *Mindset: The New Psychology of Success.* Carol Dweck. Ballantine Books 2007.
57. *Modeling, Analyzing, and Engineering NASA's Safety Culture.* Nancy Leveson 2005.
58. *Nightmare Pipeline Failures: Fantasy Planning, Black Swans and Integrity Management.* Andrew Hopkins. Aspen Publishers 2014.
59. *Normal Accidents: Living with High-Risk Technologies.* Charles Perrow. Princeton University Press 1989.
60. *Nudge: Improving Decisions About Health, Wealth, and Happiness.* Penguin Books 2009.
61. *Organizational Accidents Revisited.* James Reason. CRC Press 2016.
62. *Organizational Culture and Leadership.* Edgar Schein. Josey-Bass 2016.
63. *Organizational Learning at NASA: The Challenger and Columbia Accidents.* Julianne Mahler. Georgetown University Press 2009.
64. *Organizing for High Reliability: Processes of Collective Mindfulness: Karl Weick.* Kathleen Sutcliffe and Obstfeld, D. (1999). Research into Organizational Behavior, 21: 23–81.
65. *Patient Safety: A Human Factors Approach.* Sidney Dekker. CRC Press 2011.
66. *Performance Coaching: A Complete Guide to Best Practice Coaching and Training.* 2nd Ed. Carol Wilson. Kogan Page 2014.
67. *Pre-Accident Investigations: An Introduction to Organizational Safety.* Todd Conklin. Ashgate 2012.
68. *Presence: Bringing Your Boldest Self to Your Biggest Challenges.* Amy Cuddy. Little, Brown Spark Publishing 2018.
69. *Principles of Scientific Management.* Frederick Taylor. Originally published by Harper and Brothers 1919.
70. *Quiet Leadership: Six Steps to Transforming Performance at Work.* David Rock. Harper Business 2007.

71. *Quiet: The Power of Introverts in a World that Can't Stop Talking.* Susan Cain. Broadway Books 2013.
72. *Real Risk: Human Discerning and Risk.* Robert Long. Scotoma Press 2014.
73. *Resilience Engineering in Practice: A Guidebook.* Erik Hollnagel. Ashgate 2013.
74. *Resilience Engineering: Concepts and precepts,* Erik Hollnagel. David Woods. Ashgate 2006.
75. *Rethink: The Surprising History of New Ideas.* Steven Poole. Random House 2016.
76. *Risk Intelligence: How to live with uncertainty.* Dylan Evans. Atlantic Books 2012.
77. *Safety - I and Safety – II: The Past and Future of Safety Management.* Erik Hollnagel. Ashgate 2014.
78. Safety Culture in Nuclear Installations (IAEA-TECDOC-1329) Dec 2002.
79. *Safety Culture: Theory, Method and Improvement.* Stian Antonsen. Ashgate 2012.
80. *Safety Differently: Human Factors for a New Era.* 2nd Ed. Sidney Dekker. Ashgate 2015.
81. *Safety II in Practice: Developing the Resilience Potentials.* Erik Hollnagel. Routledge 2017.
82. *Safety, Culture and Risk: The Organisational Causes of Disasters.* Andrew Hopkins. McPherson's Printing Group 2005.
83. *Second Victim; Error, Guilt, Trauma and Resilience.* Sidney Dekker., Routledge 2013.
84. Seeing More and Seeing Differently: Sensemaking, Mindfulness, and the Workarts. Daved Barry and Stefan Meisiek. *Organizational Studies* 31(12) 2010.
85. *Sensemaking and Learning in Complex Organizations: George Tovstiga, Stefan Odenthal, Stephan Goerner.* The Fifth Conference on Organizational Knowledge, Learning and Capabilities Innsbruck 2004.
86. *Sensemaking in Organizations: Taking Stock and Moving Forward.* Sally Maitlis and Marlys Christianson. The Academy of Management Annals 2014 Vol 8. No. 1.
87. *Simple: Conquering the Crisis of Complexity.* Alan Siegel and Irene Etzkorn. Random House 2013.
88. *Simplicity in Safety Investigations: A Practitioner's Guide to Applying Safety Science.* Ian Long. Routledge 2017.
89. *Start with Why: How Great Leaders Inspire Everyone to Take Action.* Simon Sinek. Penguin 2011.
90. *Sully: The Untold Story Behind the Miracle on the Hudson.* Chesley B. Sullenberger III. Harper Collins 2009.
91. *System Failure: Why Governments must Learn to Think Differently.* Jake Chapman. Demos Publishing.
92. *Systematic Biases in Group Decision-making: Implications for Patient Safety.* Russell Mannion and Carl Thompson. *International Journal for Quality in Health Care.* Volume 26, Number 6. 2014.
93. *Tackling Risk: A Field Guide to Risk and Learning.* Rob Long and Ray Fitzgerald. Scotoma Press 2017.
94. *The Challenger Launch Decision: Risky Technology, Culture, and Deviance at NASA.* Diane Vaughan. University of Chicago Press 2016.
95. *The Checklist Mentality: How to Get Things Right.* Atul Gawande. Metropolitan Books 2010.
96. *The Death of Expertise: The Campaign against Established Knowledge and Why It Matters.* Tom Nichols. Oxford University Press 2017.

97. *The ETTO Principle – Efficiency – Thoroughness Trade-Off: Why Things That Go Right Sometimes Go Wrong*. Erik Hollnagel. Ashgate 2009.
98. *The Feeling of Risk: New Perspectives on Risk Perception*. Paul Slovic. Routledge 2010.
99. *The Field Guide to Understanding 'Human Error'*. 3rd ed. Sidney Dekker. Ashgate 2014.
100. *The Fifth Discipline: The Art & Practice of the Learning Organization*. Peter Senge. Doubleday 2006.
101. *The Five Dysfunctions of a Team. Patrick: A leadership Fable*. M. Lencioni. Wiley 2002.
102. *The Human Contribution: Unsafe Acts, Accidents and Heroic Recoveries*. James Reason. CRC Press 2008.
103. *The Invisible Gorilla: And Other Ways Our Intuition Deceives Us*. Christopher Chabris and Daniel Simons. Harper Collins 2010.
104. *The Paradox of Sensemaking in Organizational Analysis*. Organization, Florence Allard-Poesi. SAGE Publications 2005, 19 (2).
105. *The Practice of Adaptive Leadership: Tools and Tactics for Changing Your Organization and the World*. Ronald Heifetz. Harvard Business Press 2009.
106. *The Prosperous Coach: Increase Income and Impact for You and Your Clients*. Steve Chandler and Rich Litvin. Maurice Bassett Publishing 2016.
107. *The Relationship Factor in Safety Leadership: Achieving Success through Employee Engagement*. Rosa Antonia Carillo. Routledge 2020.
108. *The Safety Anarchist: Relying on Human Expertise and Innovation, Reducing Bureaucracy and Compliance*. Sidney Dekker. Routledge 2017.
109. *The Second Machine Age: Work, Practices, and Prosperity in a Time of Brilliant Technologies*. Erik Brynjolfsson and Andrew McAfee. W.W. Norton & Company 2014.
110. *The Tao of Coaching: Boost Your Effectiveness at Work by Inspiring and Developing Those Around You*. Max Landsberg. Profile Books 2003.
111. *The Wise Advocate: The Inner Voice of Strategic Leadership*. Art Kleiner, Jeffrey Schwartz, and Josie Thomson. Columbia Business School 2019.
112. *Thinking, Fast and Slow*. Daniel Kahneman. Allen Lane (Penguin) 2011.
113. *Trapping Safety into Rules: How Desirable or Avoidable is Proceduralization?* Corinne Bieder and Mathilde Bourrier. Ashgate 2013.
114. *Turn the Ship Around: A True Story of Turning Followers into Leaders*. L. David Marquet. Penguin Random House 2013.
115. *Vehicle Feedback and Driver Situational Awareness*. Guy Walker, Neville Stanton, Paul Salmon. CRC Press 2018.
116. *Why Should Anyone Be a Led By You?: What it Takes to be an Authentic Leader*. Rob Goff and Gareth Jones. Harvard Business School Press 2006.

Index

Printed in the United States
by Baker & Taylor Publisher Services

Printed in the United States
by Baker & Taylor Publisher Services